William Edward Hearn

Plutology

The Theory of the Efforts to Satisfy Human Wants

William Edward Hearn

Plutology

The Theory of the Efforts to Satisfy Human Wants

ISBN/EAN: 9783337267308

Printed in Europe, USA, Canada, Australia, Japan

Cover: Foto ©berggeist007 / pixelio.de

More available books at **www.hansebooks.com**

PLUTOLOGY:

OR THE

THEORY OF THE EFFORTS TO SATISFY HUMAN WANTS.

BY

WILLIAM EDWARD HEARN, LL.D.
PROFESSOR OF HISTORY AND POLITICAL ECONOMY IN THE UNIVERSITY
OF MELBOURNE.

London:
MACMILLAN AND CO.
MELBOURNE: GEORGE ROBERTSON.
1864.

The Right of Translation is reserved.

CONTENTS.

	PAGE
INTRODUCTION 1	

CHAPTER I. *Of Human Wants.*

§ 1. Wants common to man with other animals 12
2. Man's peculiar power of providing for these wants 13
3. Consequent expansion in man of the primary wants 14
4. Permanence of this expansion 15
5. Wants peculiar to man 16
6. Variety of human desires 17
7. Cause of this variety 18
8. Verification of the foregoing principles 20
9. Error of the ascetic doctrine respecting wants 20
10. Moral function of wants 21
11. The subject of this inquiry is the means of human enjoyment irrespective of their character 22

CHAPTER II. *Of the Instruments by which Wants are Satisfied.*

§ 1. The instruments for satisfying wants are Natural Agents and Labour 24
2. These instruments separately as means of gratification... ... 26
3. These instruments conjointly as means of gratification... ... 28

CHAPTER III. *Of the Circumstances on which the Efficiency of Labour Depends.*

§ 1. The efficiency of labour depends upon its quantity, its quality, and its direction 30
2. Circumstances affecting the quantity and the quality of labour. The natural powers of the labourer 31
3. His skill in his occupation 32
4. His general cultivation 34
5. His habitual energy 36

		PAGE
6.	The health of the labourer as a condition of energy	37
7.	His success as such a condition	40
8.	His security of enjoying the result of his work as such a condition	43
9.	Physiological principles on which the conditions of energetic labour depend	44
10.	Climate as a condition of habit of energy	45
11.	Personal habits of the labourer as such a condition	47
12.	The character of the occupation as such a condition	48
13.	Circumstances affecting the direction of labour	50

CHAPTER IV. *Of the Means by which the Efficiency of Labour is Increased.*

§ 1.	The efficiency of labour is increased by the education of the labourer, by the increase of his habitual energy and by the advance of Science and of Art	53
2.	The education of the labourer as a means of increasing the efficiency of his labour	54
3.	Examples of the economic loss arising from want of education	58
4.	The habitual energy of the labourer promoted by his improved sanitary condition	60
5.	And by increased motives to exertion	66
6.	Increase of knowledge as a means of increasing the efficiency of labour	68

CHAPTER V. *Of the Circumstances on which the Efficiency of Natural Agents Depends.*

§ 1.	Circumstances which affect the productive capacity of a country	71
2.	Its soil	73
3.	Its climate	74
4.	The presence in it of useful plants and animals	76
5.	The presence in it of natural agents subsidiary to production	77
6.	The presence in it of natural agents that facilitate intercourse	78
7.	Conditions of water-transit as illustrative of facile communication	80
8.	Negative advantages of a country	84
9.	Conditions of land-transit as illustrative of such advantages	85
10.	The physical advantages of a country depend upon the concurrence of its positive and its negative advantages	89
11.	All such advantages are relative to the existing powers of the labourer	92

CONTENTS. vii

CHAPTER VI. *Of the Means by which the Efficiency of Natural Agents is Increased.*

		PAGE
§ 1.	The efficiency of natural agents is increased by discovery	95
2.	And by improvements	99
3.	And by the increased capacity of men to use these agents	102
4.	Man proceeds from forces of an inferior to the forces of a superior potency	104
5.	Moral influence of this principle	105
6.	Land forms no exception to it	107
7.	Evidence in support of this view as to land	110
8.	How far the Ricardian theory of rent agrees with this principle	113
9.	Examination of the Law of Diminishing Returns	115
10.	This law is not peculiar to land	117
11.	Why the tendency to diminishing returns is not practically felt	118

CHAPTER VII. *Of the Aids to Labour.*

§ 1.	The aids to human powers	120
2.	Capital as such an aid	121
3.	Invention as such an aid	122
4.	Co-operation as such an aid	123
5.	Exchange as such an aid	125
6.	Practical influence of these aids	127
7.	Prevalence of errors as to their character and effects	129

CHAPTER VIII. *Of Capital.*

§ 1.	Nature of capital	134
2.	Its commencement undiscoverable	135
3.	Amount of the assistance that it renders	137
4.	Mode in which capital assists industry	139
5.	Capital is continually consumed and reproduced	140
6.	Accumulation is not necessarily a good	141
7.	General predominance of savingness	143
8.	Moral influence of capital	144

CHAPTER IX. *Of the Circumstances which Determine the Extent of Capital.*

§ 1.	The amount of capital depends upon the quantity of commodities at the disposal of the capitalist, his inclination to save, and his inclination to invest	147
2.	Capital is limited by the productive powers of industry	148
3.	The conditions of the effective desire to save are the state of the intellect, and the state of the sympathies	148

		PAGE
4.	The state of the intellect as determining the amount of accumulation ...	149
5.	The state of the sympathies as determining the amount of accumulation ...	152
6.	The conditions of the effective desire to invest are the adequacy of the consideration and the security of enjoyment ...	157
7.	The supposed adequacy of the expected advantage as a condition of the effective desire to invest ...	160
8.	Security for enjoyment as a condition of the effective desire to invest ...	160
9.	The use of capital, how limited ...	164

Chapter X. *Of Invention.*

§ 1.	Nature of invention ...	167
2.	Amount of the assistance that it renders ...	169
3.	The increase of power obtained by invention	172
4.	The saving of time so obtained ...	174
5.	The certainty of result so obtained	176
6.	The economy of material so obtained	178
7.	The power of using better materials so obtained	181
8.	Moral influence of invention	183

Chapter XI. *Of the Circumstances which Determine the Extent of Invention.*

§ 1.	Extent of invention, how limited	186
2.	Existing state of science as a limit of invention	187
3.	Practical ingenuity as such a limit	189
4.	Manual dexterity as such a limit ...	190
5.	Existing state of other arts as such a limit ...	192
6.	Inventions dependent upon the demand for them	193
7.	Some consequences of these principles	195
8.	The use of invention, how limited	198

Chapter XII. *Co-operation.*

§ 1.	Nature of co-operation ...	200
2.	Amount of the assistance that it renders ...	204
3.	Economy of effort obtained by co-operation	207
4.	Increase of skill so obtained	208
5.	Organization of work so obtained	211
6.	Saving of time so obtained	212
7.	Terms upon which co-operation takes place	213

CHAPTER XIII. *Circumstances which Determine the Extent of Co-operation.*

		PAGE
§ 1.	Conditions of successful co-operation	215
2.	Sufficiency of supply of workers	216
3.	Constancy of such supply	217
4.	Reliability of workers	218
5.	Docility of workers	220
6.	Honesty of workers	222
7.	Extent to which co-operation may be carried	224
8.	Cases where co-operation tends to excess	227
9.	Joint-stock companies	229

CHAPTER XIV. *Exchange.*

§ 1.	Importance of exchange	235
2.	Amount of its assistance	237
3.	Nature of its assistance	238
4.	Conditions of exchangeability	240
5.	Terms on which exchange is effected	244
6.	Causes of variation in price	247
7.	Effect upon consumption of permanent change of cost	251

CHAPTER XV. *Of the Circumstances which Determine the Extent of Exchange.*

§ 1.	Circumstances on which the power of exchanging depends	254
2.	Productivity of industry as a limit of exchange	255
3.	This industry is general and not merely agricultural	256
4.	Diversity of production as a limit of exchange	258
5.	Willingness to deal as such a limit	260
6.	The facility of finding purchasers as such a limit	262
7.	The facility of reaching purchasers as such a limit	263
8.	Extent to which exchange spontaneously proceeds	265

CHAPTER XVI. *Of the Reciprocal Influence of the Industrial Aids.*

§ 1.	Influence of capital upon invention	267
2.	Influence of capital upon co-operation	269
3.	Influence of capital upon exchange	270
4.	Influence of invention upon capital	271
5.	Influence of invention upon co-operation	272

		PAGE
6.	Influence of invention upon exchange	273
7.	Influence of co-operation upon capital	275
8.	Influence of co-operation upon invention	278
9.	Influence of co-operation upon exchange	280
10.	Influence of exchange upon capital	283
11.	Influence of exchange upon invention	286
12.	Influence of exchange upon co-operation	288

CHAPTER XVII. *Of the Industrial Organization of Society.*

§ 1.	The social state essential to the full development of industry	291
2.	Growth in society of tacit co-operation	293
3.	Advantages thus obtained	295
4.	Certainty with which the supply of services meets the demand	299
5.	Succession and duration of such supply	302
6.	Localization of industry in districts	305
7.	Causes which determine such localization	307
8.	Localization of industry in nations	310
9.	Causes which determine such localization	311

CHAPTER XVIII. *Of the Adjustment in Society of the Terms of Co-operation.*

§ 1.	Adjustment of terms in cases of tacit co-operation	315
2.	Difficulties incident to such adjustment	317
3.	The use of natural agents is always gratuitous	318
4.	Circumstances which obscure such gratuity	320
5.	Land forms no exception to these principles	322
6.	Why waste land in new countries brings a price	323
7.	Shares of labourer and of capitalist in expressed co-operation	325
8.	How their relations are modified in tacit co-operation	328
9.	Question of real wages, why not discussed	330

CHAPTER XIX. *Of Competition.*

§ 1.	Law of price under competition	332
2.	Nature of competition	334
3.	Its tendency is towards equality of remuneration for equal services	335
4.	Character of its operation	337
5.	Competition diffuses our gains	339
6.	And diffuses our losses	340
7.	And secures the best administration for industrial enterprises	343
8.	And checks unsuitable projects	345

CHAPTER XX. *Of the Social Contrivances to Promote Organization.*

		PAGE
§ 1.	Secondary industrial aids	348
2.	The standard of quantity	349
3.	The standard of value	351
4.	Coined money	353
5.	The prerogative of coinage	354
6.	Expedients for economizing coin	355
7.	Fairs and markets	357
8.	The distributing class	359
9.	Notification of readiness to trade	363
10.	System of public communications	367
11.	The carrying business	368
12.	Means of locomotion	374
13.	Slow progress of roads	376
14.	Railways	377
15.	Analogy between roads and money	379

CHAPTER XXI. *Of the Industrial Evolution of Society.*

§ 1.	Phenomena of organic evolution	382
2.	Similar phenomena observable in society	384
3.	Conditions and limits of organic growth	385
4.	Conditions and limits of social growth	389
5.	How man lightens the pressure of the check upon his increase	391
6.	Increase of population on reduction of the prudential check	394
7.	Principle of development in organisms and in societies	397
8.	Historical examples of evolution	398
9.	Interdependence of parts in organisms and in societies	402

CHAPTER XXII. *Of the Assistance Rendered to Industry by Government.*

§ 1.	Man is a political animal	406
2.	The primary object of political association is the maintenance of rights	408
3.	Political organization is essential to industry	410
4.	Duties of the state other than the maintenance of rights	413
5.	Matters of common interest the sole subjects of state interference	415
6.	Some applications of this principle	417

CHAPTER XXIII. *Of the Impediments Presented to Industry by Government.*

		PAGE
§ 1.	The non-performance by government of its proper functions	424
2.	The mal-performance by government of its proper functions	426
3.	Impediments from uncertain taxation	427
4.	Impediments from excessive taxation	428
5.	Impediments from raising revenue otherwise than by taxation	429
6.	Impediments from indirect taxation	430
7.	General principles of public revenue	433
8.	Impediments from the regulation of industry by the state	435
9.	Impediments from the industrial enterprises of the state	438
10.	Tendency towards the performance of the state of improper functions still vigorous	442

CHAPTER XXIV. *Of Some causes of Poverty.*

§ 1.	Positive causes of poverty	447
2.	Failures from accident or external interference	448
3.	Failures incidental to the organization of society	450
4.	Facility with which losses are repaired	455
5.	Feebleness of desire as a cause of poverty	458
6.	Preference of other objects to wealth as a cause of poverty	461
7.	Degradation of industry as a cause of poverty	463
8.	Waste of natural agents as a cause of poverty	466
9.	Existence of a criminal class as a cause of poverty	468
10.	The causes of poverty evanescent	472

INTRODUCTION.

The name Political Economy, as applied to a department of social philosophy, has been frequently the subject of unfavourable criticism. Its etymology, to some extent, involves at least an inconsistency. Its meaning is by no means distinct to an uninitiated ear. One portion of the expression relates to the Family; the other portion relates to something quite different from the Family, the State. The name hardly conveys without explanation the idea usually attached to it by economists, namely, the theory of national wealth. But no trifling defect of etymology or of precision would of itself be a sufficient reason for abandoning the use of a term which long and continued usage had sanctioned. If Political Economy be generally understood to mean the science of wealth, and if there be no other objection to its use, a writer upon that subject ought not to depart from the established practice of his predecessors.

There are, however, other objections to the use of the title Political Economy far more serious than any which the etymologist or the lexicographer can suggest. The meaning of the expression, if it be taken strictly, is such a management of the affairs of a State as is analogous to the prudent management of the affairs of an individual. In this sense Political Economy is clearly an art. It ought to lay down practical precepts for the administration of public affairs. This meaning, too, suggests an analogy which most economists now regard as false, although probably when the name was given, it was considered peculiarly exact. It seems to indicate that the Government ought to direct and control the industry of its subjects in the same manner as a great millowner or contractor now directs and controls the industry of his millhands or his navvies. In a less strict sense Political Economy may mean that management which the State, as such, should exercise for the common benefit. In this view, as in the former, we still have to do with an art. Political Economy would thus, according to either explanation, be equivalent to the art of government: in the one case, to the art of paternal government; in the other case, to the art of government generally, without reference to any particular system.

Such a meaning is not what Political Economy, if it be a science, ought to express. The art of government is doubtless of the highest importance, but it is a different thing from the theory of wealth. This theory, indeed, must, in common with many other theories, furnish some of the premises upon which that art is constructed; but it is only one out of many sciences, which, in the case of this, as of every scientific art, are thus laid under a willing contribution to the purposes of life. The distinction is itself abundantly clear; but words, as Bacon has observed, shoot back, like the Tartar bowman, upon the understanding from which they proceed: and so Political Economy, which was at first considered as synonymous with administrative art, even still, when its name has been limited to one of the sciences on which that art is based, retains in practice its original force. Thus, partly from its connotation, and partly from the nature of the subject with which it deals, the name inevitably suggests a code of practical precepts alterable by human will for the purposes of supposed convenience, and not the investigation and the statement of certain sequences that spring from the nature of man and his relations to surrounding objects.

Political Economy, indeed, has rarely, except in name, been ever regarded as a science. It is as an art that the older economists have always treated it: it is as an art that Adam Smith defines the term: it is as an art that it has acquired its present popularity: it is as an art that even those who insist upon its scientific rank habitually discuss it. In the Wealth of Nations the theoretical part is merely introductory to the practical part. Mr. J. S. Mill avowedly follows the example of Smith. Mr. McCulloch declares his opinion that a merely scientific treatise can find no readers in England. Mr. Laing insists that every country has a Political Economy of its own, suitable to its own physical circumstances and its own national character. Even Mr. Senior, who has more steadily than most economists kept before him the scientific character of Political Economy, frequently diverges into those practical applications of his theory that most occupied the public mind at the period at which he wrote. In the French and the American writers, too, as in M. Bastiat and Mr. Carey, we find a continual reference to the polemics of the day, or to the policy, or the institutions of their countries. In most of the detached essays and lectures of the present generation of economists, the subject, where it is not a discussion upon logical method, is almost always some question of immediate practical interest. Recent events have greatly increased this tendency. In the days of Malthus and Ricardo, economists fit audience found, though few: but the great success of that commercial policy which economic writers so strongly advocated, and with which the names of many of them were identified, both led to the belief that Political Economy was essentially an art, and enlisted upon its side many enthusiastic but ill-instructed admirers. Under the influence of these injudicious friends and in their impatient activity, the economic inquiries that bring fruit seem likely to supersede the inquiries that bring light. The art is built up before its foundations in the science are securely laid. It is therefore no matter for surprise, however much it may be for regret, that the progress of economic knowledge has been much slower than its friends in the earlier part of the century had anticipated.

I have said that the name Political Economy does not convey to the novice any definite meaning. Unfortunately, it is not much more distinct to the professed Economist. I do not now speak of any disputes as to the precise terms in which its formal definition should be expressed, or as to its position as a Science or as an Art. Irrespective of its relation to other sciences or of its mode of treatment, the extent of the subject is still unsettled. Apart from those extensions of the term which most economists would now regard as abusive, Mr. Senior* attributes to it three distinct meanings. He states that Political Economy may be defined, either as the science which states the laws regulating the production and distribution of wealth, so far as they depend on the action of the human mind : or as the art which points out the institutions and habits most conducive to the production and accumulation of wealth : or as the art which points out the institutions and habits most conducive to that production accumulation and distribution of wealth which is most favourable to the happiness of mankind. According to this authority, and it is one which ranks deservedly high, a writer upon Political Economy may reasonably adopt,† at his

---
* *Four Introductory Lectures on Political Economy*, p. 36.
† Ib., p. 46.

discretion, any one of these widely differing definitions. Mr. Senior admits and sanctions a still wider latitude of construction. Even when he selects the more restricted definition and proposes to deal with the subject as a science, he declares that he is entitled to discuss its applications, and to estimate, although by way only of episode, its influence not only upon wealth, but upon well being. I cannot but think that a name which admits and invites such doubts and digressions is objectionable on other than mere æsthetic grounds. Value itself is hardly a more misleading and dangerous term than is Political Economy.

There is yet another objection to this name. The former objections have chiefly had reference to the word Economy: the present difficulty is occasioned by the adjective which qualifies that word. I venture to think that, for scientific purposes, an inquiry into public or national wealth, before the principles of private or individual wealth have been ascertained, is an inversion of the natural arrangement. The wealth of a nation, in the sense in which economists use the phrase, is the aggregate of the wealth of all the individuals that compose that nation: an inquiry, therefore, into the nature of wealth and its processes in the individual ought to precede any inquiry of the same kind as to communities. The influence of the aggregation of men in a community materially affects indeed the power of acquiring wealth in each of its individual members; but this influence is rather a modification of the circumstances of the individual, than an independent subject of inquiry. Society is not different from its component parts: it only establishes between these parts certain relations which are additional to the powers of the isolated man, and do not supersede them. Political Economy, therefore, is not a proper description of the theory of wealth: it both restricts the extent of that theory, and it inverts its method; it limits the inquiry to that part of the theory of wealth which relates to the economic influence of society; and it commences the study not in its simpler but in its most complex forms.

The notion which Political Economy is usually intended to convey is expressed by the greatest expositor of the science as an Inquiry into the nature and causes of the wealth of nations. This expression, notwithstanding the great authority on which it rests, has never become popular. For our present purpose it is exposed to some of the objections which affect the title Political Economy. Both titles direct attention to the society, and not to the pre-existing parts of which that society is composed. But the expression, "the wealth of nations," presents another and a peculiar difficulty. The term wealth is both surrounded with a haze of bewildering associations, and is by no means free from ambiguity. It appears to be generally restricted to material objects; and by most economists it is expressly so limited. The reasons for which I am unable to agree with this limitation of the subject matter of the science will appear in the following pages. But wealth has also been taken as equivalent to exchangeable objects, and some of the best English economists have even proposed to call Political Economy the Science of Exchanges, or some equivalent name. I hope hereafter to show the true position which, in my view, Exchange holds in relation to the theory of wealth. For the present it is enough that I am now dealing with the means that men use to satisfy their wants. But as all exchangeable objects, by reason of their being exchangeable, satisfy some want, it follows that some objects which satisfy wants are exchangeable, and that exchange is but one means out of several of procuring us satisfaction. Neither exchange, therefore, nor wealth when used as equivalent to exchange, would suitably describe the present essay. In making, then, an attempt to inquire, without any direct practical reference, into the theory of wealth, whether the owner of it be, or be not, associated with other men, I have, but with all the timidity that becomes an innovating Briton, ventured, on the authority of a distinguished French* writer, to describe this work by the somewhat novel title of Plutology. I now proceed to explain more fully the meaning which by this term I desire to convey, and to state the scope and design of the present treatise.

Man is by his nature so constituted as to be liable to various wants and desires. He is impelled to gratify them partly by the pleasure which attends the gratification, and partly by the pain which is the consequence of its neglect. These desires depend upon man's mental and moral powers, and, consequently, expand with the expansion of his faculties. Like all other human affairs, wants and desires may be regarded from different aspects. That with which we are now concerned is neither physical nor moral, but industrial. These wants or desires when so considered form one of the chief sources of human action. They involve for their satisfaction a certain effort either personal or vicarious. These efforts, and the contrivances by which either their result is increased or their amount diminished, and the circumstances favourable or unfavourable to such contrivances, form the subject of this inquiry.

The instruments by which our desires are satisfied are either human or physical, including, under the latter term, both animate and inanimate forces. These instruments may act either separately or in conjunction. In either case the circumstances must be separately considered on which the efficiency of each instrument depends. With respect to labour, its efficiency depends upon its amount, its kind,

---

* *Traité Théorique et Pratique d'Économie Politique*, par J.–G. Courcelle Seneuil. Tome i. Partie *Théorique ou Ploutologie*. Paris, 1858.

and the direction which it receives. With regard to natural agents, they too vary in their quantity and their kind; but their utility to man is also affected by another consideration. They must be regarded not merely with reference to their intrinsic qualities, but with reference to man's ability to use them. Thus the agent which in one state of circumstances is useless or even noxious, may in other circumstances be highly beneficial. Man, then, may increase his control over nature by discovering new forces; or by applying improved methods to forces with which he is already familiar: but the principal source of his increased control is found in his own increased knowledge and increased industrial powers. This knowledge and these powers are constantly expanding; and men, consequently, are enabled to use natural agents of constantly increasing potency. Hence the course of human industry is from the worse to the better, and not otherwise. Our tendency is towards wealth, and not towards poverty; and the conditions on which the realization of that tendency depends, are our attentive study of nature and our fitting preparations to fulfil her requirements.

Satisfaction implies effort, but efforts are from their very nature more or less disagreeable. Consequently, men always strive either to reduce the amount of any particular effort, or to obtain from the same amount a larger quantity of satisfaction. With this object, four great classes of contrivances have been adopted in aid of labour. These industrial auxiliaries are Capital, Invention, Co-operation, and Exchange. Each of these has its own conditions, and its own limits; each admits of its appropriate development; and each in its turn affects the action of all the others. These agencies and these contrivances are of universal application. It is by these means and by no other that every man supplies his wants, and supplies them to the greatest advantage. Their consideration comprises the theory of wealth, so far as an isolated man, or rather an isolated family of men, is concerned. But men never live in absolute isolation; and seldom, and then only in an abnormal state, in mere groups of families. Man, partly from a sympathetic impulse, analogous perhaps to the instinct of gregarious animals, and partly from a deliberate choice, constantly tends towards the formation of that union of families which is called society. The social state re-acts most powerfully upon the industrial condition of the labourer. It largely extends the influence of all those industrial aids which exist indeed but in a merely rudimentary form apart from society. But, in addition to this expansion of the aids there is spontaneously formed that industrial organization of society itself, which forms one of the most striking phenomena that nature, amid all her wonders, has presented to us. We cannot, it has been truly said,* conceive of a more marvellous spectacle in the whole range of natural phenomena, than the regular and constant convergence of an innumerable multitude of human beings, each possessing a distinct, and, in a certain degree, independent existence, and yet incessantly disposed, amid all their discordance of talent and character, to concur in many ways in the same general development, without concert and even consciousness on the part of most of them, who believe that they are merely following their personal impulses. This natural evolution of society, which, in the absence of interruption, proceeds with all the regularity of physical development, is not limited exclusively to occupations. It further tends by localizing industry to extend its organizing influence both to entire localities in the same community, and even to different communities themselves. Like all other organisms, the social organism is subject to

---

* Comte's *Positive Philosophy* (Miss Martineau's Ed. vol. ii. p. 140).

fixed conditions and limits of growth. The same phenomena which characterize organic growth may also be observed in society. In the social no less than in the individual body, we may trace the continual augmentation of the mass, the increasing complexity of the structure, the closer interdependence of the organized parts, and the consequent advance of vital activity. But the co-operative form, which, under these laws, society spontaneously assumes, involves other important considerations. Questions arise as to the mutual rights of the several partners, which, complicated as similar questions often are even in the simple cases of ordinary partnership, become in society, and with its growth, altogether unmanageable by human legislation. These questions relate both to labour and to its auxiliaries and to the use of natural agents. It thus becomes necessary to consider the principles of wages and of profits, and to ascertain the nature and exact doctrine of rent. The machinery by which the difficulties of these questions are solved, and by which individual interests and the utmost freedom of individual action are harmonized with the general good and the permanent benefit of mankind, is found in the great agency of competition. As the development of society proceeds, various secondary contrivances are devised for increasing still further the productive powers of industry. Since it is by the extension of co-operation and of exchange that this development is chiefly manifested, those contrivances are designed to promote these objects. They facilitate the settlement of the terms upon which exchanges can be effected: and they furnish means for finding the locality of a suitable market; and for reaching it when it has been found.

But, as men are not gregarious but political animals, as they live not merely in society but in society regulated by law, we must consider, so far at least as it affects industry, that further phase of social evolution which is implied in the term Government. The influence of government upon industry is felt both in its acts and in its forbearances. The proper exercise of the legitimate functions of government is at all times most beneficial to industry, and is essential to its complete development. The improper or the defective performance of these functions, and their undue extension, are among the most formidable obstacles with which industry has had to contend.

Such is the outline which, in the following pages, I have attempted to sketch. I have desired only to state the general conditions which at all times and in all circumstances determine the efficacy of those attempts which men make to satisfy their wants. Whether in any particular case it be expedient or inexpedient for any person or for any nation to forego, either partially or altogether, any enjoyment, is a question with which I have no present concern, and from the discussion of which I have designedly abstained. The chief present impediment, as I think, to the advance of the science is that "overhasty and unseasonable eagerness" with which, in despite of Bacon's* memorable warning, economic writers "almost always turn aside to practice." This error, as the same great thinker remarks, arises partly from the desire of the material results of their inquiries; partly from an impatience resulting apparently from a want of faith in their principles; and partly from a weak craving for present popularity, less for themselves than for their subject. But whatever may be its cause, this confusion of function brings with it at the present time the same consequences as it produced in the days of Bacon. "Like Atalanta, they go aside to pick up the golden apple: but meanwhile they interrupt their course and let the victory escape them."

---

* *Novum Organum.* Aph. lxx.

CHAPTER I.

OF HUMAN WANTS.

§ 1. Life in every form with which we are acquainted, is subject to waste and repair. The living structure in no case continues unchanged, but is maintained by a series of reparative acts. If any of these acts be discontinued, life ceases, and the organism quickly disappears. In the case of animal life, provision is made by the agency of pleasure and pain for securing the proper supply of reparative material. Every animal is possessed of sensibility; and the acquisition of those materials which are necessary to keep in activity its vital powers is attended with pleasure; while the privation of them involves an equally distinct pain. Food, drink, air, and warmth, are the most urgent of these necessities. If these or any of them are withheld beyond a certain small degree or a certain brief time, the animal must die. These necessities man shares with all other animals. He must have a constant supply of pure air; he must have a sufficiency of such food and drink as his organs can assimilate. In colder climates at least, since nature has not furnished him with the protection that the lower animals enjoy, he must have more ample means than they require of retaining the vital heat. If any of these essential conditions be unfulfilled, the human animal like any other animal must die. If they be but partially fulfilled, his powers whether muscular or nervous are proportionately feeble. If he have complied with all these conditions of his existence, these powers are in a proper state for their due exercise. The satisfaction, therefore, of his primary appetites is imperative upon man. Of all his wants, they are the first in the degree of their intensity; and in the order of time they are the first which he attempts to gratify.

§ 2. But while the superior organism thus possesses all the desires that belong to the inferior, it has also by virtue of that superiority many more. Man has not only the mere animal faculties and their corresponding wants: he has also beyond all other creatures other faculties, which, beside their own requirements, seriously affect the gratification of the primary appetites. For man is able not merely to satisfy his primary wants, but to devise means for their better and more complete gratification. The food of the dog or of the horse of our time is, except where it has been modified by man, the same as that of the dog or the horse a thousand years ago. The bee constructs its cell, the spider spins its web, the beaver builds its dam, with neither greater nor less skill than that with which bees and spiders and beavers in all known times have worked. In the quality of their work, in the kind of material they employ, in the modes in which they deal with those materials, there is no improvement and there is no decline. Man alone, of all known animals, exhibits any such improvement. He alone has cooked his food. He alone has infused his drink. He alone has discovered new kinds of food or drink. He alone has improved the construction of his dwelling, and has provided for its ventilation. He alone clothes his body, and varies that clothing according to the changes of the temperature, or his own ideas of decoration. He alone is not content with the mere satisfaction, in whatever manner, of his physical wants, but exercises a selection as to the mode of their satisfaction. So strong in him is this tendency to the adaptation of his means, that, in favourable circumstances, he regards the preparation of the objects which are intended for his gratification as of hardly less importance than the gratification itself. Thus the comparative range of human wants is rapidly increased. When the question of degree is admitted in the satisfaction of the primary appetites, and when the greater or less adaptability of various objects to satisfy these appetites is recognized, the extent of human desires is bounded only by the extent of human skill.

§ 3. As the attempt to satisfy the primary appetites thus gives rise to new desires, so the actual increase of these desires tends of itself to a still further development. The enjoyment that a man has once received he generally desires to renew. The mere repetition soon becomes a reason for its further repetition. By the powerful influence of habit the desire becomes a taste; and the taste quickly passes into an absolute want. Nor is this all. The mere exercise of the faculties strengthens them ; and gives rise to a comparison of results and a desire for further improvement. The man whose senses are educated to a certain point, who has had to a certain extent experience of different modes of satisfying his desires, and has formed a judgment upon the comparative efficiency of these modes, will seldom in favourable circumstances stop at that point. Not merely would a return to what pleased his untaught faculties be intolerable to him ; but the actual enjoyment which he derives from his discovery stimulates him to further advances, and suggests the modes of obtaining them. Thus while man is not guided and limited by a blind instinct but each individual is left free to rise or fall according to the exercise of his powers, provision is made even in the primary wants of our nature both to prevent the retrogression of the species, and to secure its advancement. The number of wants that belong to this class is therefore limited, as I have said, by our knowledge of the properties of matter or of material objects fitted to satisfy our wants, and by our skill in their adaptation. This knowledge and this skill continually increases; and as the limit they present recedes, the range of our tastes and of our artificial wants increases with them.

§ 4. These principles may be readily verified. It needs no elaborate proof to show that men constantly desire an increase of physical comforts; that when they have acquired such comforts they are pained at their loss, but that their acquisition does not prevent them from continuing to desire a further increase. The universal experience of mankind is conclusive on these points. We feed and clothe and lodge our felons in a way that, to an Australian blackfellow, would seem an unspeakable luxury. The mechanic that daily complains of his hard lot would be shocked if he were reduced to use no better light, or no more convenient measure of time, than that by which Alfred wrote and by which he distributed his labours. Two pounds of tea were presented to Charles II. as a present worthy of a king. A century afterwards, the steady perseverance of the Americans in abstaining from their unjustly taxed tea was rightly regarded as the most remarkable case of national self-denial that history records. Tobacco was unknown to our ancestors, and even now is unused by not a few; yet its deprivation was in the eyes of the Irish pauper the most cruel aggravation of work-house constraint.

"It is a phenomenon, says Bastiat,* well worthy of re-

---

* *Harmonies of Political Economy*, p. 52, (English Translation).

mark, how quickly, by continuous satisfaction, what was at first only a vague desire, quickly becomes a taste, and what was only a taste is transformed into a want, and even a want of the most imperious kind. Look at that rude artizan : accustomed to poor fare, plain clothing, indifferent lodging, he imagines he would be the happiest of men and would have no farther desires, if he could but reach the step of the ladder immediately above him. He is astonished that those who have already reached it should still torment themselves as they do. At length comes the modest fortune he has dreamt of, and then he is happy—very happy—for a few days. For soon he becomes familiar with his new situation, and by degrees he ceases to feel his fancied happiness. With indifference he puts on the fine clothing after which he sighed. He has got into a new circle, he associates with other companions, he drinks of another cup, he aspires to another step, and if he ever turns his reflections upon himself, he feels that if his fortune has changed, his soul remains the same, and is still an inexhaustible spring of new desires."

§ 5. There are other important respects in which human wants differ from those of the inferior animals. In addition to those primary appetites which he shares with the humblest living creature, and which relate exclusively to things, man has also in a peculiar degree affections which relate to persons ; and various desires which are only conceivable with reference to abstractions, and result not from any physical antecedent, but from operations of the mind. By the aid of memory, which recalls the past; and of imagination, which represents the distant, the absent, and the future ; and of reason, which exercises a judgment upon the utility present or prospective of an object, and upon the means of obtaining it, man forms desires concerning his personal safety, his family, and his property. These desires like those already described become by the force of habit, daily more persistent and intense. To this class of desires no limit can be assigned other than the mental powers of each individual.

These wants, except those relating to the family, might arise in a man isolated from all other beings of the same kind. But man is by the constitution of his nature a social being. Beginning with the family he soon forms relations with other men; and lives, and moves, and has his being, in society. Hence arise new desires; each of which, like every other desire, is intensified and confirmed by habit. Man is imitative; and so seeks to have what his neighbor enjoys: he is vain; and so desires to display himself and his possessions with advantage before his fellows: he loves superiority; and so seeks to show something that others have not: he dreads inferiority; and so seeks to possess what others also possess. Hence it is that, as daily experience teaches us, no man ever attains the state in which he has no wish ungratified. The greater the development of the mental and moral faculties, the greater will be the number of desires; the more continuous the gratification of these desires, the more confirmed will be the habit.

§ 6. Human desires are indefinite not only as to their extent, but as to their objects. The capacity of desire is strengthened, and extended by exercise; but the desire is not necessarily felt for the same things. There are some objects to the use of which strict physical limits are set. There are others of which the pleasure depends, in a great degree, upon their scarcity. But in hardly any case does the increase of the object bring with it a proportionate increase of enjoyment. The sameness soon palls upon the taste; and if, as is usually the case, an extraordinary quantity of one object involve a corresponding diminution in the supply of others, one faculty or class of faculties is gratified to the full extent that its nature will bear; while the other faculties are left unsupplied.

Not merely is the amount of human desire indefinite, but the modes in which desire in many different individuals is manifested, are equally without any practical limit. Even in the primary appetites, there is room for great diversity, according to differences of climate, age, sex, and other considerations, in the choice of food, and the construction of houses, and the fashion of clothes. In the desires which are peculiar to man, we seldom find agreement. The diversity of individual tastes is proverbial. Two persons will often regard with very different feelings the same object. The same man will at different times and in different circumstances experience great changes in his desires and his aversions. There is, however, a remarkable distinction in the facility with which desires can be appeased. It is in those cases in which the commodity is essential to our existence or our comfort that the limit to our gratification is soonest reached. Our most irrepressible appetites are the most quickly satisfied. Our most insatiable desires are the most easily repressed. Were it otherwise, with the present predominance of the self-regarding affections, the accumulation of the wealthy might interfere with the existence of the poor. Desire too is never transformed into a want, strictly so called, that is into painful desire, until it has been made such by habit; in other words, until the means of satisfying the desire have been found and placed irrevocably within our reach.*

§ 7. It is not difficult to perceive the cause of this diversity of desire; or to trace the circumstances on which the development of our wants depends. That cause is found,

---

* Bastiat *Harmonies*, p. 58.

where at first it might not be expected but where its presence is consistent with a deeper investigation of our nature, in the state of our intellectual development. Beyond the mere primary appetites, no other want can make itself known, except through some mental operation. Our actions depend upon our will; and our will depends upon our judgment. If we seek to obtain any object, it is because we desire it; if we desire it, it is because we have formed some notion of its nature, and some judgment upon its suitability to our purposes. According, then, to the degree with which we are acquainted with external objects, and to the power that we possess of judging of their relations to ourselves, and to other things, our capacity of desire will be extended. Our desires, too, are subject to our will; and admit of being repressed or encouraged without assignable limits. It therefore depends upon the education, in the widest sense of that term, of each individual, and upon his character as mainly resulting from that education, how many and what kind of objects, and with what degree of persistency he desires. The more complete the intellectual development, the wider will be the field of desire; and, by the usual reaction in our mental nature, the wider the field of desire, the stronger will be the inducements to intellectual effort for the continuance of means to gratify these desires. On the contrary, the narrower our field of thought, the more contracted and the more humble will be our desires; and the less consequently will be the inducement to incur that continuous exertion of mind or body that industry implies. Where intelligence therefore prevails, the number of desires and the power of satisfying them will be alike great; where intelligence is small, the number of desires and the power of satisfying them will also be small. If this principle be true of individuals taken separately, it will not cease to be true of them when they are regarded as forming the aggregate that we term a nation.

§ 8. It requires but little observation to perceive the confirmation which these reasonings obtain from actual experience. We know that the desires of educated men are more varied and more extended than those of persons without education. We know that the wages of educated men are higher, and consequently their means of gratifying their desires greater, than those of the uneducated. If an educated man be reduced by misfortune, we sympathize with the disproportion between his desires and his means of satisfying them. If an uneducated man become suddenly rich, we see that, from the limited extent of his former wants, and the undeveloped condition of his desires, he literally does not know what to do with his money, and rushes into the most extravagant and ludicrous follies. We see that if a man be content, like a dog, to eat his dinner and to sleep, his nature will gradually sink to that of the brute. The higher faculties will waste from disuse; the lower in the absence of restraint, and from habitual exercise, will acquire a complete predominance. On the other hand, those nations, and those classes of a nation who stand highest in the scale of civilization, are those whose wants, as experience shows us, are the most numerous; and whose efforts to satisfy those wants are the most unceasing.

§ 9. Nothing, therefore, can be farther from the truth than the ascetic doctrine of the paucity and the brevity of human wants. So far from man wanting little here below, his wants are indefinite; and never cease to be so during his whole existence. Nor is there anything immoral in such a view. The supposed inconsistency arises from a confusion of apathy with content. The former term implies that the development of desire is repressed ; the latter that it is regulated. Content is a judgment that, upon the whole, we cannot with our existing means improve our position, along with an unmurmuring submission to the hardships, if any, of that position. Its aim is, not to satisfy desires ; but to appease complaint. It is consequently not inconsistent with the most active efforts to alter that combination of circumstances upon which the judgment was formed. "The desire of amelioration, it has been* truly said, is not less a moral principle than patience under afflictions ; and the use of content is not to destroy, but to regulate and direct it."

§ 10. So far from our wants being unworthy of our higher nature, we can readily trace their moral function and appreciate its importance. They not only prevent our retrogression, but secure our advancement. Our real state of nature consists not in the repression, but in the full development and satisfaction, of all those faculties of which our nature consists. Such a state is found, not in the poverty of the naked savage ; but in the wealth of the civilized man. It is the constant and powerful impulse of our varied and insatiable desires, that urges us to avoid the one state, and to tend towards the other. "Wants and enjoyments," says Bentham,† "these universal agents in society, after having raised the first ears of corn, will by degrees erect the granaries of abundance, always increasing, and always full. Desires extend themselves with the means of gratification ; the horizon is enlarged in proportion as we advance, and each new want, equally accompanied by its pleasure and its pain, becomes a new principle of action. Opulence, which is only a comparative term, does not arrest this movement when once it is begun ; on the contrary, the greater the

---

* Dr. W. C. Taylor's *Natural History of Society*, vol. i. p. 145.

† *Works*, vol. i. p. 304.

means, the greater the field of operations, the greater the reward, and consequently the greater the force of the motive which actuates the mind. But in what does the wealth of society consist, if not in the total of the wealth of the individuals composing it ? And what more is required than the force of those natural motives, for carrying the increase of wealth to the highest possible degree ?"

But these wants do not stimulate our acquisitive and inventive powers only. They also serve to discipline our moral nature. Many of man's proceedings are slow in their nature ; and so he must practise patience. In like manner, he must expend some of his acquisitions with the view of acquiring more ; and thus in addition to patience he must exercise hope. One great means of increasing his power is co-operation with his fellow-men : he must therefore, to some extent, subordinate or at least assimilate his will to theirs ; and so he must learn forbearance. Thus the efforts that we make for the satisfaction of our wants supply the means for developing both our intellectual and our moral faculties.

§ 11. The subject of this inquiry is the efforts made by man to secure enjoyment. The particular character of any enjoyable object is, therefore, for the present purpose indifferent. The question is not whether a given object be conducive to our general well-being ; but simply whether it be enjoyable. If it be enjoyable, it is foreign to the purpose to consider whether the enjoyment to which it contributes be unmeaning or even immoral ; or whether it be embodied in a tangible shape ; or be merely a fleeting gratification of the sense; or be a permanent benefit to the body or the mind. We pass no judgment upon the character of the want, or upon the manner in which it should be regulated. For our purposes, wants are simply motives of varying power which universally exist, and the laws of which we propose to investigate. We have to deal with them merely as forces, without any other estimate of their characters than the intensity with which they are felt by the persons who experience them. Nor are we any more concerned to appreciate the character of the means of enjoyment, than we are to appreciate the character of the want. It is enough that the want is felt, and that it can be satisfied.

CHAPTER II.

OF THE INSTRUMENTS BY WHICH WANTS ARE SATISFIED.

§ 1. There are some natural agents, which, without any conscious action on the part of man, supply his most urgent wants. The air, the light and heat of the sun, the force of gravity, affect him irrespective altogether of his will. But these are rather conditions precedent of existence, than objects of desires, which by any care of his own man can satisfy. In every case, other than that of mere involuntary action, a perceptible effort must intervene between the desire and its gratification. The will must be consciously directed towards the attainment of the object; and force, whether muscular or nervous, must be brought into action accordingly. But every such effort, as the very term seems to imply, is more or less troublesome. It has, indeed, been contended that labour at least in moderation is in itself agreeable; and that man is attracted to industry by the pleasure which the exercise of his faculties for a determinate object affords him. The spontaneous activity of the child, or the habitual exertion of the disciplined worker, are indeed pleasurable; but the case is very different with the ordinary class of efforts. The activity of the child, when it has served its destined purpose, gradually decreases with his growth, just as the pleasure in any new undertaking disappears with the novelty; and its place is supplied by the compulsion which education in a greater or less degree involves. A man often feels pleasure in his work; but only after years of training, and the formation of that habit which proverbially amounts to a second nature. But in every country, and in every age, uncivilized, or partially civilized, men delight in inactivity. In all the acquired energy of civilized life, it is for rest from labour that the active man sighs: it is to such rest that, untaught by all the experience around him, he looks for the fulfilment of his Elysian dreams. It is with no small difficulty, and after much training, that the youth's natural aversion to continuous exertion, whether of mind or of body, is overcome. We may, therefore, safely conclude that it is not less true that men desire abundance, than it is that they dislike exertion.

It follows then that man does not seek the gratification merely of his desires. He seeks their gratification, at the least possible sacrifice of his comfort. The accomplishment of his wishes, and the economy of his efforts, are the two conflicting conditions which he incessantly attempts to reconcile. Thus at the commencement of our inquiries respecting wealth, two questions present themselves. One of them asks by what means men satisfy their wants. The other asks by what contrivances this satisfaction is most easily effected. To each of these questions, in due order, the present chapter and its successors will attempt to furnish a reply.

Desire consists in a certain state either of body or of mind. Its satisfaction arises from the presentation of an appropriate object to the part so disposed. This presentation, whether it be made by the person himself who feels the desire, or by some other person in his behalf, must be made by some human agent. The object so presented may be supplied either from some physical source, or from the exercise of some human faculty. Thus human energy may be exerted with or without reference to external objects. There are, consequently, two, and only two, instruments by which human wants are satisfied. These instruments are natural agents and labour. In the former term I include all forces, whether animate or inanimate, other than man. By the latter term I mean the exercise of any human faculties for a definite object. Each of these agents, either separately or in conjunction with the other, may supply some want or accomplish some purpose of man.

§ 2. There are some natural agents which, in the ordinary sense of the term, need no labour. They are available for human purposes at once, without any change of form or of place, and with no further effort than that of mere appropriation. In some favoured parts of the earth, the bounty of nature is so profuse as, even with regard to man's proper wants, to dispense almost, if not altogether, with the necessity for exertion. In those happy climes, in which the dreams of poets are almost realized, the earth produces spontaneously her choicest fruits, and the genial sky demands no shelter. The rice, the palm tree, or the cocoa nut, supply a profusion of nutritious food: the temperature renders all but the scantiest covering superfluous: and the shade of a tree, or a wigwam of branches affords a sufficient protection against the gentle dews of night. But these cases are exceptional; and the wants thus supplied are merely man's primary needs. There is no instance in which nature has done for man more than this: there are few in which she has done so much.

Independently of any material object, man possesses certain faculties, the exercise of which gives pleasure to himself or to others. He can give advice: he can afford protection : he can relieve pain. He can please the ears with his singing, or the eyes with his painting. He can narrate, the past: he can describe new scenes of fancy : he can present to the senses his brilliant dreams with all the vividness of actual life. His constant attendance, even his mere society, is desired.

It has been usual to distinguish all those kinds of labour which do not result in the production of a material commodity from those kinds which have such a result ; and to designate the latter kind productive, and the former unproductive. If these epithets be, as they are popularly understood to be, laudatory or the opposite, we should have not only to determine what is enjoyable ; but to distinguish between what is useful and what is pernicious. If, as the best writers limit them, they relate merely to the tangible or the non-tangible character of the result, we should exclude from our consideration not the least numerous or least important part of every civilized community. The distinction seems to have arisen from regarding a part only of the subject as if it were the whole. Material commodities form one species of the means for satisfying human wants: they are indeed an important but still they are only a single species. Men are not all engaged, nor are all those who are so engaged exclusively concerned, in growing corn or printing calicos. All the great professions, all the liberal arts, the execution of the laws, the administration of justice, the cultivation of literature, the pursuits of science, may surely be regarded as contributing in their several ways to human convenience or enjoyment : yet all these occupations and many more, if wealth, in the ordinary sense of the term, be alone deserving of consideration, must either be entirely shut out from the science, or must be forced into it by some unnatural distortion. Every writer may of course at his discretion select some particular part of a subject ; and may carefully abstain from transgressing the limits he has proposed for himself. If, therefore, a writer select for his subject those efforts to procure human enjoyment which result in the production of material objects, he is justified in omitting all those other efforts which do not produce a tangible result: but it should at the same time be understood that the view which he undertakes to give of the question is partial and incomplete; and that there are other sources of gratification not less important than material objects, and not less susceptible of scientific treatment.

§ 3. But, although there are thus cases in which natural agents alone and labour alone are capable of satisfying our desires, the influence of these agencies when separate is insignificant in comparison with their influence when combined. By far the greatest portion of our wants depend for their satisfaction upon some combination of natural agents and of labour. Sometimes labour modifies the form of natural agents: sometimes natural agents increase the powers of labour: generally both these processes are combined, and natural agents are modified by labour, assisted by other natural forces. In this case there are two agents, the presence of both of which is essential. The production of commodities requires men to work and material to be worked. Man, indeed, can only combine or alter the materials that he finds. He can direct the agencies around him: the properties of matter do the rest. On the other hand, almost all natural objects admit of being rendered suitable to satisfy human desires or to accomplish human purposes; but they are seldom fit to do so in their original state. Those again that are fit, do not readily present themselves in the form or at the time or at the place at which they are required; but can only be obtained after a tedious and labourious search. Coal exists in many countries in abundance; but it is hundreds of feet below the surface of the earth. Oil is unlimited in its possible supply; but the whales that yield it are swimming amid polar seas. Guano may be obtained in some places for the trouble of taking it; but it must be sought over distant seas, and conveyed in costly ships. If we choose, we may ourselves mine for our coal, or sail in search of our oil, or our guano. If we prefer it, we may for a suitable consideration induce some other person to incur the trouble on our behalf. But in any case the trouble must be taken. Labour is the purchase money of all things. Even in Paradise man was placed in the garden to till it. We are surrounded by objects suited to gratify our wants; but we must ourselves put forth our hands to gather them.

Labour, then, whether by itself or as operating upon appropriate external objects, may be practically regarded as the primary condition of the gratification of our desires. It is the agent which, in the language of Adam Smith, "originally supplies every nation with all the necessaries and conveniences of life which it annually consumes." I proceed accordingly to consider the circumstances on which the efficiency of this great instrument depends; and the means by which this efficiency can be increased. The next inquiry must be into the external conditions under which it works; or, in other words, the presence and the character of those natural agents, upon or in connection with which labour is usually exerted. We have then to investigate the auxiliaries by which its power is increased, and industrial success thereby secured.

CHAPTER III.

OF THE CIRCUMSTANCES ON WHICH THE EFFICIENCY OF LABOUR DEPEND.

§ 1. The efficiency of labour, as of all other things, depends upon the quantity of it that the labourer can command and upon the character of that quantity. It is not however easy in the case of labour to keep separate these conditions. Not merely is an alteration in quality in many cases equivalent to an increase or a diminution in quantity; but the same causes affect alike both quantity and quality. It will therefore be convenient to treat both branches of the subject together, without attempting to draw any marked distinction. But there is a further condition peculiar to labour. In the case of natural agents, considered in themselves and without reference to man's capacity to use them, we are concerned only with their amount and the kind of their force. Their operation is the same in whatever manner and towards whatever object they are directed. The process of digestion goes on whether the subject presented to the stomach be food or poison. The powder will explode and the bullet be projected whether the gun be directed towards the person himself who pulls the trigger or towards the object which he desired to strike. But as every action is voluntary, and as consequently the direction of our faculties may be altered at our pleasure, not only must the amount or the quality of the faculties exercised be considered but also the judgment with which that exercise takes place. The greatest amount of the best kind of labour, if it be improperly applied, may be wasted; while exertions in every other respect inferior may, under proper direction, produce considerable results. Since therefore human action proceeds from the will, and thus differs from the blind external forces which we guide, our estimate of labour must include the circumstances which affect not only its amount and its quality but also its direction.

§ 2. The efficiency of labour depends in the first instance upon the powers of mind and body that the labourer possesses. Although it is difficult to distinguish between what we have received from nature and what we owe to art, wide differences, both in physical and mental power, obviously exist between man and man. These differences may be regarded as merely individual peculiarities; and disappear when we have to deal with masses. The excesses of some and the deficiences of others are mutually balanced. It is probable that between any two equal masses of people belonging to the same nation, the average result of their respective natural powers would show but little difference. Between men of different countries, and still more of different races, however, the result would not be so uniform. Whether they be the effect of artificial advantages during several generations, or some more latent cause, differences do in such cases exist, and to a sufficient degree to form an important economic consideration. Experiments,* although perhaps not wholly satisfactory, have been conducted on this subject. Their result is said to prove

---

* See Quetelet on *Man* (Chambers' Ed.) pp. 68, 113.

that French sailors are at least one-third stronger than Australian aborigines and that natives of the British Isles, and especially Irishmen, surpass in muscular power the inhabitants of the Low Countries. It has been found that two English labourers or artizans are equal in productive power to three Danes, or three Norwegians, or three Swedes, or three Norman labourers, or three Germans.* Between Englishmen and men of a different race, such as the Croats or the Hindoos, the disparity is considerably greater. The French and the Italians are unequalled in those arts which require delicacy of touch and refinement of taste. The Russians are said to possess considerable inventive and imitative powers, and to show a remarkable aptitude for industrial and commercial pursuits.† The Finnish races on the contrary, although they occupy the same country as the Russians and mingle freely with them, are still without trade. The Jews have at all times shown a singular talent for money dealing. The Armenians and the Bokharians are unrivalled pedlers.‡ But as the actual character of the labourer is more important than the fertility or the abundance of the natural agents that surround him; so the circumstances that form his character are more important than his mere undeveloped power and capacities. Some of these circumstances will presently appear. It is sufficient now to indicate the fact that between different individuals of the same nation and between different nations considerable diversities of original power exist: and that such diversities, other things being equal, determine, and in all cases concur in determining, the efficiency of labour.

§ 3. Next to the natural powers of the labourer, we

---

* *Journal of Statistical Society*, vol. xxv. p. 509.
† Haxthausen, *Russian Empire*, vol. i. pp. 152, 156.
‡ Ib. vol. ii. p. 32.

have to consider his acquired powers. Of these the most
important are his peculiar skill and aptitude in performing
the particular work or service that he undertakes. The
senses become by constant exercise acute to an almost incredible degree. Astronomical observers can estimate differences of time to the tenth part of a second, and differences
of space to the five thousandth part of an inch. Operators
in the electric telegraph recognize the peculiar voices
of their several wires. A skilful artizan will in
a few minutes complete a work which would be
entirely beyond the powers of a person unacquainted with
that art. A professional man every day, almost without
conscious effort, disposes of matters which, to an unprofes-
sional person, seem hopelessly perplexed. Even between an
experienced practitioner and a novice in the same art with
equal or even considerably superior natural endowments, the
difference in the efficiency of their services is frequently very
great. The actual perfection of skill has a tendency to weaken
our appreciation of it. Men are often inclined to under-
value the importance of a service when they see the apparent
ease with which it can be rendered. They forget the time
and labour and many failures by which that facility of action
has been attained. Those athletic feats which seem the
easiest, and are most readily performed, are in reality the
results of the most careful training. The same rule, and
probably for the same reason, holds good in other matters
also. The economy of effort is no easy lesson to learn.
It takes a long time to understand how much exertion,
whether of mind or body, is enough, and not more than
enough, to accomplish any given object. It takes a still
longer time to acquire the habit which is essential to propriety and ease in making that exertion. There are many
feats, both of body and of mind, that the untrained worker
is quite unable to perform. There are many others through

which, with difficulty indeed and awkwardly, he can go. But it is the perfection and the test of true skill to conceal all traces of effort; and to accomplish with apparent ease and quickness the work which, in the hands of another, could only be produced, if at all, after long toil and in a far inferior shape.

§ 4. The influence of skill in augmenting the efficiency of labour in any special branch, whether of bodily or of mental labour, is sufficiently obvious. But the economic advantage of general intelligence, though not less true, is not equally manifest. It was formerly, indeed, the received opinion that instruction in anything beyond their mere art tended to unfit men, at least those in the humbler ranks of life, for the duties of their calling. Such an opinion would at the present day find, at least in our country, few supporters. The improvements that have recently been made in every kind of instrument require skilled workmen for their use ; and a workman can in no other way so readily become skilled as by a preliminary general training. It is said that artillery recruits are now sent for some time to the regimental school before commencing their military instructions. The acknowledged quickness of the volunteers in learning their evolutions and other military duties, as compared with regular troops, has been explained by the superiority of their general education. Those military officers who have obtained their commissions by competitive examinations, are said to learn their drill in half the time that those appointed under the old system of patronage would require.* Men of great experience in the employment of labourers† on the continent of Europe have declared their preference for the Saxons and the Swiss, over the workmen of all other countries. The ground of this

---

* *Journal Statistical Society*, vol. xxii. p. 66.
† See Mill's *Political Economy*, vol. i. p. 133.

preference is that the Saxons and the Swiss "have had a very careful general education, which has extended their capacities beyond any special employment, and rendered them fit to take up, after a short preparation, any employment to which they may be called." The superior efficiency of American, over immigrant, operatives which arises from the superior intelligence and better education of the former, is estimated by an able American author as equivalent to twenty per cent.* The same writer states that the Board of Education in Massachusetts procured from the overseers of factories in that state a return of the different rates of wages paid, and of the degree of education of those who received them. From this return it appeared that the scale began with those foreigners who made a mark as the signature to their weekly receipts for wages, and ran to the girls who taught in school in the winter months, and worked in the factories in the summer.

"It may be asserted," to quote the words of a celebrated economist,† "that the degree, not of any particular instruction, but of common, general instruction, of which a nation is capable, may be taken as the measure of its civilization. It is by instruction that the human mind is brightened; that it opens itself to intellectual pleasures; that its tastes are refined; and that man experiences the want of ennobling, and, if I may so express myself, of purifying his enjoyments. Instruction causes him to experience new wants other than the gross wants: he wishes to procure the means of satisfying them; and thus his natural activity is stimulated. Instruction scatters those fatal prejudices which, regarded merely from an economic point of view, are opposed to the development of social wealth. Instruction, in fine, gives

---

* E. P. Smith's *Manual of Political Economy*, p. 107.
† Rossi, *Cours d'Économie Politique*, tome 4, p. 71.

more foresight, more prudence, as well to those who produce as to those who consume."

§ 5. But however great may be the natural powers of the labourer, or however consummate his skill, or however bright his general intelligence, the industrial importance of these qualities manifestly depends upon the mode in which they are exercised. It is not the mere existence of natural or acquired powers, but their actual employment that determines their utility. The principal regulator, therefore, of the efficiency of labour, is the habitual energy with which the labourer pursues his work. It is not enough that a man should on an emergency be capable of making great exertions. Such fitful efforts are generally followed by a corresponding reaction; and at best fall far short of the effects of steady and constant work. It has been often observed that savages are capable at times of great exertion and great endurance; yet their intermittent efforts do not even bear comparison with the steady and continuous industry of civilized men. In every occupation, we daily see the success which attends patient perseverance; and the comparative failure of even great natural powers when irregularly exerted. The clever workman who wastes half the week in idleness or dissipation, but who in the remaining half can earn what is sufficient for his support, is gradually left behind by his less quick but more persevering competitor. Similar results are familiar in professional life. Instructors of youth know well the advantages of the regular discipline and the daily work of a school, over the thousand interruptions that occur in home education. Equally marked is the difference between the results of the mere cram for an examination, and of the regular and careful preparation that goes on from day to day and year to year. The fable of the hare and the tortoise is indeed an oft repeated tale.

But while the work must be habitual, it must also be energetic. Many men pass their lives over their work, with far less results than a more resolute and vigorous spirit could accomplish in a mere fraction of the time. The "government stroke" is now a proverbial expression. The imperturbable *insouciance* of a Dutch or German tradesman is to an Englishman or an American almost distracting. "Look," says Mr. Laing,* "at an Englishman at his work, and at one of these Dutchmen, or any other European man. It is no exaggeration to say that one million of our working men do more work in a twelvemonth, act more, think more, get through more, produce more, live more as active beings in this world, than any three millions in Europe in the same space of time; and in this sense I hold it to be no vulgar exaggeration that the Englishman is equal to three or four men of any other country. Transplant these men to England; and under the same impulse to exertion, and the same expeditious working habits which quicken the English working class, they also would exceed their countrymen at home in productiveness. It is not in the human animal, but in the circumstances in which he is placed, that this most important element of national prosperity, this general habit of quick, energetic, persevering activity resides."

§ 6. The above quotation from Mr. Laing suggests that the habitual energy of the labourer, although the importance of the subject seemed to require a separate notice, is itself the result of more general causes. Energy depends upon the health of the workman, and upon the motives that induce his exertion. The frequency with which he exercises that energy, depends upon the absence of any counter-

---

* *Notes of a Traveller* (Ed. 1854), p. 8.

acting influences. These influences arise from the climate, from the personal habits of the workman, and from the nature of his employment. These two classes of causes must, therefore, be separately examined. Each of them materially affects the efficiency of labour, as well by its separate action, as because it contributes to produce that energy which is so important an element in industry.

The positive causes of energetic labour directly affect the quantity as well as the quality of labour. They imply that the powers of the labourer shall continue unimpeded, as well as that his inclination to exercise those powers shall have full scope. It is therefore an essential condition, not only of energetic work, but of work at all, that the worker's health be such as to admit of this exertion. Ill health operates, so far as it extends, as a suspension of the worker's powers, whether natural or acquired. The physical and mental prostration that accompanies disease, the tedious convalescence, the loss of time, and the actual outlay consequent on such a misfortune, and the interruption thus occasioned to industrial connections and habits, are only too familiar. Even those slight but frequent ailments which generally attend upon a weak constitution, seriously impair, while they last, the energy of the worker; and in the aggregate, make serious inroads upon his time. Previous to the reform in the mode of admission to the Civil Service, many sickly young men, too weak to bear the competition of an open profession, obtained official employment. "The extent," it is said,* "to which the public are consequently burdened, first with the salaries of officers who are obliged to absent themselves on account of ill health, and afterwards with their pensions when they retire on the same plea, would

---

* *Report on Permanent Civil Service*, by Sir S. Northcote and Sir C. E. Trevelyan, p. 4.

hardly be credited by those who have not had opportunities of observing the operation of the system."

In a still greater degree the efficiency of the army has been impaired by sickness. If the nominal strength of the army were two hundred thousand men, more than fourteen thousand trained soldiers, each representing an outlay of £100, would be habitually in hospital in time of peace.* In the Crimean War thirty-nine per cent. of our troops were sick on an average of seven months. The pecuniary loss thus caused is very apparent. The operations of the army were protracted by the prevalence of sickness: the war consequently was unnecessarily protracted; and millions of money were thrown away. It is now well understood that nothing is so expensive as an unhealthy military force.

Again, it has been ascertained that in the case of members of Provident Societies among the working classes, for every death there were not less than 465 days of sickness so severe as to lead to successful application for relief: thus, amongst these industrious and careful men, each man that dies represents a loss to the community of more than a year and a quarter of more or less productive labour, independent of the ultimate loss by avoidable death.†

The efficiency of a man's labour does not depend upon the state only of his health: it also depends upon the length of time that that state lasts. This influence is more readily perceived by considering collective than individual instances. Our attention in the case of the individual is so concentrated upon the fact of death, that we fail to think of the wealth that the deceased might have accumulated, or even of the talents and the skill that are buried in his grave. Yet the advantage of a long-lived over

---

* *Journal of Statistical Society*, vol. xxiv. p. 480.
‡ Gairdner's *Public Health*, p. 311.

a short-lived population is very great. The same period of helpless infancy, of uninstructed youth, of learning to labour, must be completed, whether the labourers live to an average of forty or of sixty. When to this simple advantage is added the value of practised skill and of experience in all occupations, it appears at once how great is the increase of wealth that arises from longevity alone. To these considerations we must add the burden cast upon society of maintaining the helpless widow and orphans of the labourer whose life has been prematurely ended, and the less efficient education which those children almost necessarily receive. Few sources of pauperism are so prolific or so painful as the lengthened illness of the head of a family or his untimely death.

§ 7. But the principal cause of energy is found in connection with the motives which induce men to work. On this point there are two fundamental conditions of energetic labour. The first is, that the labour should be productive: the second is, that the labourer should be sure of receiving the results of his toil. There is no more potent stimulus to exertion than success; and no more certain cause of relaxation than failure. We daily see how men who have attained to eminence in their several walks of life, when fresh demands upon their powers are made, rise to the occasion and perform with credit an amount of work which formerly they would themselves have thought impossible. On the other hand, men show an almost instinctive abhorrence to merely fruitless labour. In the old mythology no more terrible punishment could be devised for guilt of the deepest dye than an eternity of useless toil. Among ourselves, in prison discipline, it is not the least severe part of the punishment on the treadmill or at the crank that the men feel that all their exertions are expended merely on " grinding the wind.'·

The energy with which men work, even when after a long

depression they at length take heart and hope, is one cause of the continued success of new and prosperous communities. "The gold-fields," says a writer, describing the scenes which he witnessed in the earlier days of the alluvial diggings,* "are a scene pre-eminently calculated to exhibit the continuous powers of human bones and muscles, and a gold digger working on his own account is the personification of these powers. Few know what men can do and how willingly they do it under an adequate stimulus. We gazed at laborious and incessant industry, which neither dazzling sun nor pelting rain could cause to intermit. A number of German mining parties were met with, which had been generally successful. Little accustomed even to see gold, much less to possess it in such abundance, the peasantry of the Fatherland roused their every energy, and we heard of labors in their pits and tunnels continued by torchlight during the night as well as by light of day." Of an opposite state of things the history of Scotland presents a remarkable example. No men in the world exhibit a greater degree of habitual energy than the Scottish subjects of Queen Victoria: yet when her great-grandfather was heir to the throne the Scottish people were conspicuous for their incorrigible indolence. The lazy Scotch were, in the last century, as notorious as the lazy Irish of a later day. In both countries a like effect was produced by a like cause. The people were indolent because they were hopelessly poor. When the bulk of the agricultural population of Scotland were tenants-at-will; when their life was a constant broil with Highland caterans; when black mail was levied almost at the gates of Glasgow; when wages in the most prosperous parts of the country were fivepence a day in winter and sixpence in summer; when there were no roads and few enclosures; when the

---

* Westgarth's *Victoria*, p. 252.

whole currency of the country did not exceed two hundred thousand pounds; there was little inducement for any one to work. When these obstacles were removed; when order was maintained, and leases granted, and roads made, and lands enclosed, the laziness of the Scotch disappeared;* just as the Irish now cease to be lazy when they are placed in circumstances even moderately favourable to industry.

The perfection of unenergetic labour seems to be attained when an habitually indolent worker is engaged at low wages, through a kind of charity, upon useless public works. We may perhaps have seen among ourselves something not very unlike such a phenomenon. Still even on relief works, an English or even an Irish labourer, unless he were corrupted by a long course of idleness, would not fail to show at the end of the day some result. It is among the Italians, whose indolence under the combined influence of physical and social and political causes has been often noticed, and especially among the Romans, whom the memories of their glorious past serve but to render the idlest of the Italians, that the art of doing nothing seems to have been cultivated with unexampled success. In the year 1860 the Roman Government, in order to relieve the distress that prevailed among the poor of the city, employed several hundred men to excavate the Forum. "The sight" writes an eye-witness "of these men working, or rather pretending to work, is reckoned one of the stock jokes of the season. Six men are regularly employed in conveying a wheelbarrow filled with two spadefuls of soil. There is one man to each handle, two in front to pull·when the road rises, and one on each side to keep the barrow steady. You will see any day long files of such barrows so escorted creeping at a snail's pace to and from the Forum. It is hardly necessary to say

---

* Smiles' *Lives of the Engineers*, vol. ii. p. 95, and the authorities there cited.

that no progress whatever has been made in the excavations, or in truth is likely to be made. Yet all these workmen are able-bodied fellows, who receive two pauls a day for doing nothing."* There can hardly be a better illustration of the economic importance of energetic labour than the contrast between the Roman working on charity-wages for a government that, as he knew, would not or dared not dismiss him, and the Australian miner working on his own account in a payable claim.

§ 8. Of equal importance is that other condition of energetic labour, which requires that the labourer shall be secure of enjoying the fruits of his industry. It is to this principle that free labour owes its great superiority, when other circumstances are equal, over slave labour. So powerful is its influence that slaveholders have at all times and in all countries felt it expedient not to enforce in its full rigour the law which gave to them all the acquisitions of their slaves; and have permitted those unfortunates to retain some small portion of their own earnings. Even that mitigated form of slavery which consisted in compulsory labour for the lord during a certain number of days in the week has been found to be less advantageous than hired labour. In those parts of Europe where it still exists, it is estimated that the labour of the freeman is nearly three times as efficient as that of the serf.† As free labour surpasses compulsory labour, so the most productive form of free labour is found where this sense of security is most complete. Men like to feel that all the fruits of their labour will be their own. Thus labour by the piece is generally more efficient than labour by the day; and labour by an individual, or at least a small party, than

---

* *Journal of Statistical Society,* vol. xxiii. p 238.
† Jones on *Rent,* p. 50; and see *Journal of Statistical Society,* vol. xxv. p. 510.

labour of which the returns are shared among a considerable number. This sense of property is indeed the very breath of life to industry. All writers on the subject have described with astonishment the wonderful "magic of property" which, in the hands of peasant proprietors, have transmuted into gold the rocky valleys of the Alps, or the sterile sands of Flanders. They speak with wonder, though with perfect unanimity, of the almost superhuman industry of the peasant proprietor ; of his unwearied assiduity ; and of the affectionate interest with which he regards his land and tends every particular plant that it produces. Nor are they less unanimous in assigning the cause of such industry. Men work thus because they are working for themselves. Far different indeed are the results where the labourer has no security that he will profit by the results of his toil. Even during the present reign large districts in Ireland, under the fostering care of Receivers in the Court of Chancery, were described as presenting the appearance of a country ravaged by an invading enemy. In the greater part of that unhappy country the tenant dreaded to make the most ordinary improvement, lest he should thus afford an opportunity for an increase of his rent. There is no more certain truth than that expressed in Arthur Young's well-worn aphorism, "Give a man the secure possession of a bleak rock and he will turn it into a garden : give him a nine years' lease of a garden and he will convert it into a desert."

§ 9. The explanation of these conditions of energetic labour is found in Physiology. Labour implies muscular contraction. The contraction of the muscles depends upon the action of the nerves. Nervous excitement requires a constant stimulus. A stimulus which gives pleasure makes less demands upon nervous energy than one which causes pain. The stimulus of hope is pleasing, while that of fear is painful. Under the latter influence, therefore, there is a greater expenditure of nervous energy in the production of the same amount of muscular contraction than under the former. Hence a man will voluntarily undergo fatigue, the exaction of which would be excessive cruelty. Although the work may be the same, his actual expenditure of force is in the two cases very different. Accordingly, free men readily perform an amount and a kind of work that slaves can with great difficulty be forced to accomplish. A slave will not directly refuse to obey an order; but his labour will seldom secure the desired result. When free men are subject to compulsion, as in the case of soldiers or sailors, they exhibit a tendency towards the same vices as those which characterize slaves. If an order displease them, they obey the letter but defeat its intention. Every skilful officer is well aware of this tendency; and endeavours by appropriate means to counteract its operation. No army or no ship's company can be effective when its *morale* is depressed.

§ 10. Of the negative causes affecting energy of labour the most important is climate. This great influence affects both man's longevity and his power during his life to labour, and also his inclination to do so. In all tropical climates, at the mouth or on the banks of their great but sluggish rivers, rich masses of vegetation accumulate; and retain in deadly abundance the poisonous vegetable ooze and animal *debris* of the streams. In such countries, too, the sudden and violent changes of temperature, and an atmosphere unfavourable in its electric conditions and tainted with sulphuretted hydrogen, or some other source of impurity, increase the prevailing unhealthiness. But whatever may be the causes of disease, it is well known that there are some localities highly unfavourable to the continued duration of human life. The malaria in many parts of Italy; the yellow fever in New Orleans and other Southern towns; the pestilent shores of Western Africa; Darien, not less deadly at the present day than in the time of its first ill-fated colonizers all these, and many other instances, show the importance to labour of a healthy or unhealthy climate. The effect of climate upon the power of labouring while the health remains unimpaired is equally apparent. The enervating influence of continued heat, especially when the atmosphere is at the same time damp, and the bracing influence of a colder climate has been proved by abundant experience of residents in India and other tropical countries. In these countries it is unsafe and certainly disagreeable to engage in any active occupation during the greater part of the day. Many hours are thus lost to industry, or are expended in recruiting the demand made upon the system by the intense heat. In northern latitudes, again, external nature is so buried during many months of each year as to check the ordinary operations of industry; and the want of light, even more than the intense cold, precludes all out-of-door occupations. As we approach the limits of the temperate zone, the same results, although in a proportionately less degree, follow. Thus, the quantity of labour is diminished, and at the same time the labourers tend to become irregular and desultory in their habits. They can make for a short period the most vigorous exertions; but these efforts are invariably followed by long intervals of inactivity and repose. It has been observed* that the Swedes and Danes, and the Spaniards and Portuguese, although different races of men and with different temperaments and characters, in all other respects show with a very remarkable coincidence, both in their history as nations and in their private daily life as individuals, the

* Laing's *Denmark*, p. 204.

same tendency of character to act by fits and starts. Continuous action for long periods in the ordinary occupations of life is interrupted by frost, snow, and short daylight in the northern climes, and by extreme heat, drought, and the baked state of the soil in those of the south. Thus "alternations of great exertion and great repose are in those countries forced by climate into the life and action of every man in his out-door business and industry, and become habitual and characteristic of nations."

§ 11. Material checks, both to the energy of labour and to the habit of that energy, are also found in the personal habits of the labourer, and in the nature of his employment. There is hardly any description of personal excess that is not inconsistent with steady work; but by far the most important, both in its frequency and in its effects, is intemperance. It is needless to describe the effects of this terrible vice upon health and longevity; or its indirect influence, through its well-known tendency to produce crime, upon the direction of labour. These results are too apparent, and have been too often the subject of comment, to require in these pages any lengthened notice. But the injurious effects of intemperance are also felt both in checking the actual energy of labour, and in preventing the formation of those habits of perseverance which are even of greater importance than energy itself. Nor do these consequences of intemperance more need illustration than the others to which I have referred: they lie on the surface of our ordinary life, and are matters of daily observation. The aching head and the trembling hand refuse to yield energetic labour. The capricious refusal to work, the quarrelsome and insubordinate disposition, the disinclination of exhausted nature to exertion, and the uncertain recurrence of these feelings, gradually give rise to a degree of irregularity which is but another name for the absence of habitual energy. Whatever may be the abilities of the labourer, if there be no security that these abilities will be exercised when they are required, a great part of their utility is lost: and this unsettled and unsteady disposition must constantly re-act upon the mode of life.

§ 12. Connected with this class of cases is the difficulty which sometimes springs from the character of the employment. The nature of the labourer is often subdued to that he works in. Many occupations require exertions which are in themselves considerable, but are repeated only at intervals. Although the energy of the labour may be very great, there are thus no means of acquiring the habit of energy. An effect is produced similar to that which follows from the extremes of temperature. The principle in each case is the same: the steady course of labour is interrupted, in the one case by the climate, in the other by the intermittent nature of the demand: and this interruption, from whatever cause it proceeds, renders industry fitful and irregular. In Italy, for example, the products of the country are such as to require great numbers of hands for short periods. Wages are generally paid in kind, and as thus there is an abundance of cheap food, and as the climate requires but little shelter and is favourable to an out-of-door life, there is little inducement to regular industry. The idle listless habits of the population that result from such a state of things are very striking. "Time and labour," says Mr. Laing,* "seem not worth saving in their estimation. The women are universally sauntering about, spinning wool or flax with the distaff and spindle. A woman will spin as much yarn at her spinning-wheel in an hour as in a week with her

---

* *Notes of a Traveller* (Ed. 1854), p. 198.

distaff and spindle. But I doubt if a spinning wheel could be found in Naples. I have seen two men carrying between them, slung upon a pole on their shoulders, a common sized paving-stone: one of them could have transported six such stones in a common wheel-barrow with ease. Boats are manned with six or seven or even ten men: a man and a boy, or at the utmost two men, would be the crew of such a craft in any other country. I have seen two asses with a driver to each, and a padrone or overseer on horseback to attend them, employed in trailing into town two sticks with each ass, one on each side of the saddle, the sticks positively of a size that one of the drivers might have carried the whole four. In every job the padrone, the helper, the looker-on, the talker, and the listener, seem indispensable personages."

In our country, in certain classes of occupations, the same effects, although to a great extent neutralized by the pervading habits of industry, are clearly discernible. The coal-heavers, porters, and other persons whose labour is severe but intermittent, are seldom equal to steady and continuous work. Professional beggars appear to be quite incapable of even finishing a job. Among the semi-mendicant class, the regular labour and discipline of the Poorhouse is an effective check upon the number of the applicants for relief. It is remarkable that all these classes are more or less unthrifty. The coalheavers, for example, earn very high wages; but they are habitually in a state bordering on destitution. The same phenomena have often been observed as characteristic of savage life: there are the same irregular alternations of excessive labour and excessive repose; and there is the same extraordinary improvidence. The coincidence can hardly be accidental. It seems that mere fitful industry is not sufficient fully to develope those faculties which are requisite for the exercise of foresight.

We may trace a similar effect, though considerably less in degree, where the amount of employment is not sufficient to occupy the labourer's whole time. In this case, the irregularity arises not from the climate or the habits of the labourer or the intrinsic nature of the labour; but from the limited demand by the public for the labourer's services. But whatever may be the difference of form which the cause assumes, the fundamental principle is the same. The irregularity exists in some shape, and irregularity is inconsistent with habitual energy.

§ 13. It thus appears that the efficiency of labour, so far as regards its quantity and its quality, depends upon the powers whether natural or acquired of the labourer; and upon the energy with which he habitually exerts them. But assuming that the labour is of the greatest amount, and of the best kind that the nature of the case will admit, its efficiency further depends upon the judgment with which it is applied. I do not now speak of the utility or the inutility of the result, either to the labourer himself or to other people. I assume that as far as he is himself concerned, the labourer is the best judge of the sufficiency of the return to his labour: and that as regards others, he keeps within the limits of law. But if he consider the object worth his endeavour to obtain it, and if he avoid any injury to his neighbours, the discretion with which he applies his labour is still, with a view to his own success, a matter of essential importance. A man may attempt an object that it is impossible to attain; or he may employ means that are inappropriate, or inadequate, or extravagant. It is sufficiently plain that if we direct our labour to the production of any desirable result, that result must be capable of being produced: and further, that the efforts we make must have a tendency to produce it. Nor will it be disputed that if we apply a force capable of raising half a ton to the removal of a ton, or if we apply to the removal of a ton a force capable of moving two tons, we waste, in the one case all our efforts, in the other case, as much power as we beneficially use.

In this matter, however, our illustrations can better be obtained from the negative than from the positive side. Examples of the search after impossibilities are found in the labour of the Alchemists; in the time and trouble expended upon the attempts to square the circle; and to obtain perpetual motion; perhaps we may fairly add in the search of our ancestors for El Dorado, and in the search of our fathers and ourselves for the north-west passage. So too in mining enterprises, thousands of pounds have been expended in searching for coal, for gold, and for other minerals, in places in which a competent geologist could at once have predicted that no such mineral could exist, or if it existed, be approached. Sir John Herschel* mentions a curious instance of inappropriate means. An attempt was made to facilitate the smelting of iron by directing the current of steam from the engines that were used to drive the bellows that fanned the fire of the furnaces, at and from the boiler into the fire. It was thought that as the ingredients of steam were favourable to combustion, the force of the fire would be greatly increased. Unfortunately, however, the new arrangement instead of increasing the fire blew it out, a result, says Sir John Herschel, "which a slight consideration of the laws of chemical combination, and the state in which the ingredient elements exist in steam, would have enabled any one to predict without a trial." Of defective means, the attempts of the Phenician and Egyptian mariners to effect any extended oceanic navigation with the appliances then at their disposal

---

* *Natural Philosophy*, p. 46.

present a strong illustration. And again, of extravagant as well as inappropriate means, familiar examples may be found in the faulty organization (at least in the times before reform) of many departments of the public service.

CHAPTER IV.

OF THE MEANS BY WHICH THE EFFICIENCY OF LABOUR IS INCREASED.

§ 1. The efficiency of labour, on whatever objects it is exercised, depends in the first instance upon the natural powers of the labourer; upon the skill that he has acquired in the particular occupation in which he is engaged; and upon his general intelligence. His skill is evidently the result of experience, whether obtained by himself or communicated by others; and his natural powers, both of body and of mind, are severally capable of receiving great additional strength and extension. The first means, therefore, of increasing the efficiency of labour, is the education, in the largest sense of that term, of the labourer.

The efficiency of labour further depends upon the habitual energy of the labourer. Energy depends upon the health of the labourer, and the influence upon him of the motives to labour. These motives are success and security. Labour, therefore, becomes more effective when the circumstances which improve that health or excite those motives are increased. The uninterrupted action of the causes of energy is necessary for the formation of energetic habits. The interruptions arise principally from climate, or from the misconduct of the labourer, or from the nature of the employment. The removal, therefore, of these disturbing influences, or any of them, tends to promote the efficiency of labour. But the influences of climate, so far as they admit of remedy, are counteracted by an improved hygiene or by increased knowledge. The remedies for misconduct are included under education. The consequences incident to the nature of the employment, except so far as they are produced by removable causes, must be taken to form part of the cost that the service rendered involves. Apart then from the removal of these disturbing forces, the second positive means of increasing the efficiency of labour includes on the one hand, sanitary and hygienic improvements; on the other hand, improvements in economic arrangements and in positive law.

The efficiency of labour is also affected by the degree of judgment with which it is employed. This judgment consists in· the adaptation of suitable means to the designed ends. These ends may be impracticable; or the means may be ineffective or extravagant. A knowledge of the properties and laws of natural agents, and an ingenuity in the application of this knowledge to practical purposes, consequently modify both our objects and our attempts to attain them. The third means, therefore, of increasing the efficiency of labour, is the advancement of science and of art.

§ 2. If proof be desired of the influence of education in developing the physical or mental powers, we have but to look at the muscular condition of the athlete before and after his training; or to consider the raw school boy whom skilful and assiduous culture has developed into the accomplished scholar. Competent witnesses attest that for all ordinary civil labour four partially trained or drilled men are as efficient as five men that are undrilled. Others assert, that by early and complete training, three men may acquire the working power of five.* It would require a very great extension of education to bring the masses to the present efficiency of the best of their number, namely, foremen and non-commissioned officers. "Let any one," says Mr. Chadwick,† "who has been in a position of civil or military command, and who knows those sound and trustworthy and most excellent classes, the non-commissioned officers and foremen, estimate what that economical advance would be—a manufactory of foremen as working men—a regiment of non-commissioned officers as privates." I shall not attempt to illustrate the effect of special instruction in any particular art. Nor shall I repeat what has been already said respecting general intelligence. But all training, even that which is purely physical, involves some moral discipline. It calls forth self-denial and obedience; and it tends to produce habits of attention and of exertion for some object beyond the present. Hence the influence which teaching exercises upon the minds of those who receive it, is quite disproportionate to its extent. The religious and moral influence of public schools appears to be much greater than their intellectual influence. Experience has shewn that "a good set of schools civilizes a whole neighbourhood."‡ In the great bulk of public schools few children remain beyond the age of ten or eleven years. The amount of knowledge that a child at such an age can in the most favourable circumstances acquire, is very small; and yet even in the best schools not more than one fourth of the children reach the first class; and so never obtain a fair degree of instruction. The moral effect, however, of even this brief training seems

---

* *Journal of Statistical Society*, vol. xxv. p. 519.
† *Ib.* p. 517.
‡ *Report of Commissioners on Education in England*, p. 275.

to be indelible. " It is so much more natural to be civil, that when the habit is once formed, it is never lost." Even if a child forget all the information that he obtained at school, he has yet received a training in obedience, attention, observation, and facility of comprehension, the effects of which are visible for the remainder of his life. It is said that the demeanour of the children in the streets of a town show at once, to an experienced observer, whether the neighbourhood is, or is not, well supplied with schools; and a similar distinction has been noticed between those children who have, and those who have not, attended a well conducted infant school.*

The most decisive proof of the importance of this training, at least in the aspect in which we are now considering it, is what may be termed its commercial value. Mr. Chadwick† states that he has been at much pains to ascertain from employers the comparative efficiency of educated and uneducated labourers, and that all intelligent witnesses of wide experience and observation unanimously agree that education, even in its present rude and in many respects objectionable condition, is highly remunerative. Masters who have been at the expense of schools on high religious and social grounds, concur in saying that success is great on economical grounds. They find the readiness with which a well-educated man comprehends instructions, the willingness and the intelligence with which he makes trial of unaccustomed processes, and the quickness with which he learns them, the accuracy with which he notes the facts that come under his observation and the facility with which he reports them, the suggestions for the improvement of his business that he is able to offer, the

---

* Ib., p. 31.
† *Journal of Statistical Society*, vol. xxv. p. 516.

diminished amount of superintendence that he requires, and the saving of waste from untrustworthiness, from blundering, from misconduct, and from misdirected labour, are advantages which the mercantile mind is not slow to appreciate.

Accordingly, in the manufacturing districts of England, and especially in those districts where the business is carried on in large establishments, the employers of labour exercise a supervision over the education of the work people in their employ which is unknown elsewhere.* They contribute largely to support the schools, and sometimes compel, by weekly stoppages from their wages, all the persons in their employ to join in the contribution. Many great establishments have also incurred considerable expense in providing for their work people not merely schools, but churches, libraries, and other means of intellectual improvement. "There are," says Mr. Tremenheere, "few of the masters who would not be willing to purchase for a large sum, if that were possible, the advantage and satisfaction of surrounding themselves with a labouring population, intelligent, easily reasoned with, provident, diligent and moral."† In many cases gross neglect has been shown towards the multitudes of people that have been collected from all sides into one place; and great losses have not undeservedly befallen those who were so obtuse to their moral responsibilities, and who so improvidently risked a great capital in dependence upon ignorant and ill-conducted operatives. But where care has been taken to promote the education of the people, the results have almost invariably been beneficial. At certain extensive iron works in the county of Durham which gave employment to 15,000 people,

---

* *Report of Commissioners on Education*, p. 76.
† *Report of Inspectors of Mines*, 1850.

great pains were taken both for the personal comfort and for the general improvement of the work people. At length, under the influence of a union, a strike took place among the colliers at these works, without apparently any reasonable cause. As it met, however, with no support from the other classes of workers, it caused but little interruption to the work, and lasted not more than about six weeks. In reference to this strike the manager said, that unexpected and vexatious as it was, it showed that, even in the lowest sense of pecuniary gain, the company were amply repaid for all their expenditure upon the moral and intellectual improvement of the population. "Had they not been supported by the bulk of their people, had any ill-founded sympathy been shewn by the miners, labourers, and others, with the misguided colliers, the strike might have lasted many months instead of a few weeks, and might have obliged them to put out furnaces, and discontinue other parts of their works, a proceeding which would have cost them five times as much as they have spent in raising the intelligence, securing the comfort, and promoting the good morals and religious training of their people."*

§ 3. The evils which arise from the dullness, the prejudices, and the misconduct of an uneducated population are of long standing and have been severely felt. "It is the standing complaint," says Mr. Burton,† "of successful projectors, that they have had the greatest difficulties to combat with in the stupidity and incapacity of the persons they have had to employ; and in their inveterate prejudice against doing anything that happens to be beyond their old accustomed line of occupation. The directions which a little intelligence

---

* *Report of Inspectors of Mines*, 1850, p. 57.
† *Political Economy*, p. 276.

would enable the workers clearly to understand and fully to practise often appear, if they be out of the usual nomenclature of their trade, to be as unintelligible to them as to the lower animals. In a country like this, full of ardent, enterprising spirits, perpetually developing new shapes of enterprise, the difficulty of imparting to others a little of their own promptitude of thought and intellectual resources is a heavy interruption to progress, and leaves the elements of much wealth undeveloped." Experienced mechanicians assert that notwithstanding the progress of machinery in agriculture, "there is probably as much sound practical labour-saving invention and machinery unused, as there is used; and that it is unused solely in consequence of the ignorance and incompetency of the work-people."* Most of the lamentable accidents which occur in the application of steam power, both in manufactures and in locomotion, may be traced to the want of suitable education in the labourers. Even where machinery is employed, and without disaster, its powers can seldom be prudently used to their full extent, and it is frequently worked with lamentable waste.

A curious illustration of the loss arising from defective intelligence is found in the Portuguese vinegrowers. Portugal, as well from its geographical position as from various local advantages, possesses extraordinary facilities for the production of wine; but the farmers will study neither the best mode of cultivation nor the best methods of preparing their produce. They continue to make bad wine from good grapes for no other reason than that they and their fathers before them have always done so. "The Portuguese agriculturists," says Mr. Paget, "are incapable of imbibing the idea that nature at times requires the aid of art. They do

---

* *Journal of Statistical Society*, vol. xxv. p. 516.

as their forefathers did, disdaining the counsels of scientific men ; censuring those who attempt to introduce innovations, and more especially rejecting the advice or suggestions of those most likely to be of service to them, viz.—those who have to cater for the taste of the consumer." *

A similar example is found in Spanish history. The mine of Almaden, in Spain, produces mercury of the finest quality and in great profusion. In the early part of the last century the supply of this metal, in the face of a growing demand, was found to be constantly diminishing. At length the Spanish Government commissioned an Irish naturalist, named Bowles, to visit Almaden and ascertain the cause of the failure. Mr. Bowles found that the miners had acquired the habit of sinking their shafts perpendicularly, instead of following the direction of the lead : and reported to the Government that if a shaft were sunk obliquely the mine would again become productive. The Government ordered that this suggestion should be carried into effect. But the Spanish miners were too tenacious of their old customs to give way. They sank their shafts in the same manner as their fathers had done : and what their fathers had done must be right. At length it became necessary to take the mine out of their hands and to bring new miners from Germany. The immediate consequence of the change was, that the yield of mercury was doubled and its cost to the consumer correspondingly lowered.†

§ 4. The same argument which proves the practical influence of education upon labour is also applicable to sanitary and hygienic reform. The sound body as well as the sound mind has its commercial value. The heads of

---

* *Reports of Secretaries of Embassies and Legations*, 1858, p. 187.
† Buckle's *History of Civilization in Europe*, vol. ii. p. 107.

great manufacturing establishments have found by experience that a reliance upon unhealthy workmen is not less dangerous to the profitable application of their capital than a reliance upon ill-conducted or ignorant workmen. The employers now sufficiently understand that "it is their direct and plain pecuniary interest to take especial care of the physical energy and condition of their workpeople."* Accordingly, they frequently incur great expense in providing house accommodation and medical assistance for the persons in their employment; and in endeavouring to prevent as well as to remedy disease. We have, however, the means of estimating with tolerable accuracy the amount of the annual tribute which human ignorance, misconduct, and neglect unnecessarily pay to death. It has been ascertained that there are every year in Lancashire 14,000 deaths, and 398,000 cases of sickness which might be prevented; and of these deaths, 11,000 are those of adults engaged in productive labour. Every individual in Lancashire loses nineteen years or nearly one-half the proper term of his life; and every adult loses more than ten years of life, and from premature old age and sickness much more than that period of working ability. "Without taking into consideration," says Dr. Playfair,† "the diminution of the physical and mental energies of the survivors from sickness and other depressing causes; without estimating the loss from the substitution of young inexperienced labour for that which is skilful and productive; without including the heavy burden incident to the large amount of preventible widowhood and orphanage; without calculating the loss from the excess of births resulting from the excess of deaths, or the cost of maintenance of an infantile population, nearly

---

* *Journal of Statistical Society*, vol. xxiv. p. 463.
† *Ib.*, vol. xxiv. p. 135.

one-half of which is swept off before it attains the age of two years, and about 59 per cent. of which never become adult productive labourers; and with data in every case much below the truth; I estimate the actual pecuniary burden borne by the community in the support of removeable disease and death in Lancashire alone at the annual sum of five millions of pounds stirling."

The same causes which diminish the longevity of the labourer impair his health and weaken his powers during his life. In England the normal period of death, that is the termination of life which the natural change of our physical structure brings with it, appears to arrive at about the age of eighty: the annual death-rate ought therefore to be about twelve in each thousand.* Practically, however, this theoretical death-rate is increased by the operation of causes which in our present circumstances cannot wholly be eliminated. Partly from these causes, and partly no doubt from the influence of the same causes which in other parts of the country produce an excessive death-rate, the most healthy districts in England show a death-rate of fifteen in the thousand: but it may be stated generally that in a population living in ordinary favourable circumstances at the present day in England, the annual death-rate is about seventeen in the thousand. About a million of the inhabitants of England are living on these terms. But the death-rate is in some places as high as thirty-six, and on the average of the whole country is nearly twenty-three. All the deaths therefore beyond the lower rate are a mere waste of human life. The causes which produce this waste may be, and in many cases have been, investigated, discovered, and removed; and the quantity and quality of labour is thereby proportionately increased.

---

* Levi's *Annals of British Legislation*, vol vi. p. 57.

But if the causes of mortality are avoidable, the causes of sickness may also to a still greater degree be avoided. The usual course of consumption, which is now regarded as a preventible disease, is generally supposed to extend on an average from one to two years; and there can be no doubt that this disease involves the loss in every case of many months of valuable time. Even the most rapid and the most fatal epidemic fevers, in addition to the actual number of deaths that they occasion, cost to the community a large sum for sickness. "In the most fatal period of the most severe epidemic of typhus I ever witnessed," says* Dr. Gairdner, "about one in four of those admitted to the hospitals died. It would certainly not be too much to calculate the average loss of time in each case at more than an entire month, including the convalescence of those who recovered. For every fatal case, therefore, there must have been at least four months of serious illness."

It is, of course, difficult to convey any adequate idea of the extent to which the efficiency of labour may, by sanitary improvements, and the advance of medical skill, be increased. Some conception may be formed by a comparison of the rate of mortality at the present day with that which prevailed two centuries ago. It has been calculated from a comparison of the Tontines established under William and Mary and a century afterwards by Mr. Pitt, that a portion nearly equivalent to one fourth of the total period of existence was in the course of that century added to human life. In the Tontine of 1690, the number of years on which a man of thirty years of age might reasonably calculate, was twenty-six; in the Tontine of 1790, it was thirty-three. In the Tontine of 1690, eight persons of certain ages and sexes obtained an average existence of about fifty-six years. In the

---

* *Public Health*, p. 312.

Tontine of 1790, the same number of persons of the same ages and sexes, obtained an average of about sixty-eight years.*

So, too, in London, the average death-rate for the whole duration of the Long Parliament of Charles II., was upwards of seventy in the thousand. At the present time it is twenty-two in the thousand. In one memorable year, the plague, a disease now unknown to us, destroyed the third part of the inhabitants of London. In 1859 the number of persons who died in London was less than 62,000. If the mortality had continued at the average rate of two centuries ago, the number of such deaths would have exceeded 194,000. But the nature of disease, and the climate of London, remain the same as they were at the time of the Restoration. We must look elsewhere for the causes of this change. On this subject there is now little room for doubt. The supply of food, and especially of vegetables and of fresh animal food, was frequently deficient. The houses were as close and as dirty as they now are in the filthiest city of the East. The supply of water was imperfect; and its use was by no means general. The streets were ill-regulated: and the sewerage was defective. The soil was saturated with the accumulated refuse of many generations. The ground, on which rows of palaces now stand, was a swamp pestilent with malaria. Medical science was very imperfect; and, such as it was, was very imperfectly applied. Croup and scarletina were confounded with measles, and with fever: and the mortality from these four diseases was four times what it is at the present day. For every woman who now dies in childbirth, five were allowed to perish before the Revolution. Eight persons in every hundred thousand now die of dysen-

---

* *Transactions of the National Association for the Promotion of Social Science*, 1857, p. 501.

tery. Two centuries ago, the proportion of victims claimed by that disease was 763.*

Sierra Leone was at one time described, and with truth, as the white man's grave. At the present day, an Englishman who observes due precautions may live there as safely as in his own country. Yet the climate is the same as it was fifty years ago. The change is in the sanitary precautions adopted by the inhabitants. Experience has taught to the Europeans one of the most difficult lessons an Englishman can learn, the adaptation of their habits of life to the circumstances of their position. They now build more suitable houses than before; they dress in a manner adapted to the vicissitudes of the climate; they abstain from the excessive use of alcoholic drinks. The practice of the art of medicine has also improved, and it is said that the general use of quinine has considerably shortened the period of illness, and reduced the mortality.†

An attempt has been made to obtain a pecuniary expression for the effects of these improvements. It has been estimated that the annual money saving to the community by the improvements recently effected in the town of Liverpool, irrespective of the value to each individual of his own life, of the mental and bodily suffering that has been prevented, and of the saving of the money sunk in young lives that passed away before their labour became productive, amounts to £617,500. It has also been calculated‡ that if the prevalence of consumption alone could be reduced in England by one-half, or even one-third, the service so rendered would greatly exceed in mere money value a sum equal to the interest of the national debt.

---

* Levi's *Annals of British Legislation*, vol. xi. p. 348.
† *Journal of Statistical Society*, vol. xix. p. 62.
‡ Gairdner's *Public Health*, p. 128.

§ 5. Next after the longevity and the health of the labourer, the principal elements in determining the energy of his labour are success and security. Ample results for his exertions, and full assurance that he shall himself enjoy these results, are, as we have seen, the conditions most favourable to energetic labour. The former is evidently a reacting cause. Men labour energetically because they are successful; and their exertions are successful because they are energetic. As then the labourer is more largely assisted by the agencies to which I shall presently refer, as the impediments to his exertions are removed, as he acquires the use of improved natural agents, and applies improved processes in dealing with them, the return to his labour becomes larger; and the motive to exertion is consequently stronger. If there be any doubt on this subject, we need only compare the labour of the same man when hope has disappeared, and when his services are in full demand and meet with large remuneration.

While these improvements in production thus conduce to its success, security is ensured by improvements in the relation of the labourer to other men. Without law there is no security; without security there is no industry; without industry there is no abundance. Every improvement, therefore, either in positive law, or in the positive morality of the community, has a direct tendency to promote the efficiency of labour. The industrial history of Scotland illustrates the effects of this kind of improvement. I have already noticed the semi-barbarous state, so strangely contrasted with its present indomitable industry, in which, even in the middle of last century, that country remained; and I have indicated the causes of that condition. The people were indolent because their labour was unsuccessful; and their labour was unsuccessful mainly because the enjoyment of its fruits was insecure. The Highland caterans

plundered them from without; their landlords still more
irresistibly plundered them at home. Happily, almost at the
same time at which the tardy vigour of the Government was
aroused against the marauders, some security was obtained
against extortion under the cover of the law. A Lothian
gentleman,* Mr. Cockburn of Ormiston, observing that one
of his crofters or small tenants had enclosed his field at his
own cost, rewarded this unparalleled enterprise by a lease for
nineteen years, with a promise of renewal for a similar
term. The results were found so satisfactory that Mr.
Cockburn was induced to extend the practice; and it soon
was generally adopted throughout the country. From that
time agriculture advanced with extraordinary rapidity, and
the whole appearance of the country was entirely changed.
The practice which experience had so strongly and so quickly
recommended was soon extended by the provisions of the
Act known as the Montgomery Act. This Act, which was
passed in the year 1779, counteracted to some extent the
mischievous effects of the Scotch law of entail by enabling
the tenant in tail both to charge the inheritance for money
expended on the estate; and also to grant leases for nineteen
years, the period introduced by Mr. Cockburn, and in certain
circumstances and under certain conditions for longer terms.
In some respects, too, the law of Scotland was more favourable to the tenant than that of the sister country. In
addition to these advantages the Government did not think
it worth its while to interfere with the humble beginnings
of that system of banking which, arising spontaneously from
the state of the Scotch law of debtor and creditor and
the exigencies of Scotch society, afterwards rendered such
service to the country. Under all these influences the
agricultural industry of the Lowlands of Scotland, for it

---

* Smiles's *Lives of the Engineers*, vol. ii. p. 108.

was in the Lowlands only that these influences were in full operation,* has become the model and the school for that species of industry in all the other parts of the kingdom.

§ 6. It is self-evident, as Mr. Mill has observed,† that the productiveness of the labour of a people is limited by their knowledge of the arts of life ; and that any progress in those arts, any improved application of the objects or powers of nature to industrial uses, enables the same quantity and intensity of labour to raise a greater produce. Human knowledge and human power, according to Bacon's celebrated aphorism, meet in one. Nature to be commanded must be obeyed; and that which in contemplation is as a cause, becomes in operation as a rule. We know that when events have happened in a certain sequence if the same assemblage of circumstances be again obtained, the same sequence will again ensue. When in any particular case this sequence has been established, and the whole of its circumstances ascertained, we have practically obtained a new power. We can reproduce at our will the phenomenon. Knowledge thus gives us foresight, and foresight leads to action. As knowledge therefore increases, men can perceive what in the circumstances they should do and what they should forbear from doing. They understand the objects which they may reasonably expect to accomplish ; and the amount and the kind of labour which the attainment of these objects involves. They also understand, what is only of less importance than the knowledge of their power, the extent of their inability. They

---

* See two papers in the third volume of the *Transactions of the Dublin Statistical Society*, by W. Neilson Hancock, LL.D., entitled respectively, "What are the Causes of the Prosperous Agriculture in the Lothians of Scotland?" and "What are the Causes of the Distressed State of the Highlands of Scotland?"

† *Political Economy*, vol. i. p. 130.

know the objects which either absolutely or at any particular place or time they cannot attain; or the respects in which the means that they apply fall short of, or exceed their needs.

In the present age, the maxim that knowledge is power has been literally and amply verified by the great mechanical inventions with which we are familiar. To these, however, I do not now refer. Invention, or the substitution of inanimate forces for human action, although auxiliary to labour, is distinct from it. But apart from these, there are many remarkable examples of the extent to which human labour, alone, and unaided by new physical forces, has been enabled by increased knowledge of the properties of its subjects to increase its results. The greatest advances that have been made in agriculture have consisted, not in any mechanical inventions, but in the application of more judicious processes to the earth itself, and to the plants growing on it. Where the land was formerly left idle in every second or third year, the rotation of crops now admits of its uninterrupted cultivation. Where the powers of any land have been exhausted by excessive cropping, the application of appropriate manures will restore its fertility. Some land produces wheat but not clover: other land is good for clover, but will not yield turnips. Vast sums have been wasted in attempting, by mere experiment, to fertilize land for plants which it would not otherwise bear, and in blindly applying the experience thus blindly obtained. But the elements essential to each kind of plant have been ascertained; and their presence in any soil, or their absence from it can also be determined. In this way the causes of barrenness and the remedy for it become at once apparent. Bogs and marshes may thus be turned into cultivable land; and the arid wilderness may be made a blooming pasture. There are new modes of pruning and of training, and of supporting plants and trees. There are new principles of plantation and of the preparation of the soil. In manufactures and in commerce time may be saved, or material economised, or returns expedited. Hides, the preparation of which formerly required many months, may now be tanned in a few weeks. Large retail shops excel their smaller competitors partly by their improved methods of shop-keeping, and partly by their adherence to the principle of small profits and quick returns. In all these cases the amount of production has been increased merely by additional skill, and the use of improved processes.

CHAPTER V.

OF THE CIRCUMSTANCES ON WHICH THE EFFICIENCY OF NATURAL AGENTS DEPENDS.

§ 1. Whatever may be the degree of excellence to which our labour has attained, or however judicious may be its direction, its result, where that direction is towards a material object, must be greatly influenced by the quality of that object. The importance to the labourer of superior natural advantages is sufficiently conspicuous. Every person knows that the natural forces available for human uses which one country possesses, are often more abundant, or more accessible, or of a better kind than those in another country. In a country so favoured the same result is attained with a diminished effort as compared with the labour of less fortunate countries. Experience, indeed, has shown that natural advantages, like the advantages of fortune, do not always bring to the possessor the full benefit that might be derived from them. In the lands which owe most to the bounty of nature, the higher faculties of man have not received as full a development as they have experienced under colder skies, and on a less fruitful soil. No great nation in tropical latitudes has continued permanently free and with free institutions. No such nation has equalled in wealth, much less in good service to humanity, the inhospitable New England, or the stern and wild Scotland, or Holland with difficulty rescued from the sea. But a fertile soil, and a genial climate, and an abundance of all things useful to man, are in themselves good. Whether they prove blessings or curses to those to whom they are given, or remain unappreciated by them, depends in this case, as in all other cases, on the men themselves. But whether we use our advantages or pervert them, whether we permit them to destroy or to develope our faculties, it does not become us to convert, like the Trojan prince of old, into a matter of reproach, those "very glorious gifts of the gods, such as they, of their bounty, may give us, but of his own will none may take."*

The productive capacities of any two places are often very different. From those idle deserts where no man may dwell, to those more favoured regions which, for human purposes, are from the very superabundance of vitality hardly less unavailable, from the snows of the Polar Circle to the forests of the Amazon, the degrees of productivity are infinite. These differences arise from the differences between various countries in the quantity and the quality of their respective natural agents. We may roughly classify, for our present purpose, those forces that directly minister to human wants. Some forces, either spontaneously or in co-operation with human efforts, produce objects which are, at once and in themselves, useful to man. Others furnish instruments which facilitate such production. There are others which favour communication between different countries or different districts; and so conduce to the distribution of industrial

---

* οὔτοι ἀπόβλητ' ἐστὶ θεῶν ἐρικύδεα δῶρα,
ὅσσα κεν αὐτοὶ δῶσιν ἑκὼν δ' οὐκ ἄν τις ἕλοιτο.
Iliad, b. iii. v. 65.

products. In addition to these positive effects, a country may possess what may be called negative advantages. There are natural agents which are obstructive to human settlement; and which must be modified or removed by human exertion. The absence of these agents, therefore, forms an important consideration in determining the comparative utility of any country to man.

§ 2. The differences between countries in their capacities for direct production are due mainly to differences in their soil and their climate, or either of them ; and in the character of the plants and animals that they respectively contain. Of the combined influence of soil and climate, the tundras of Northern Asia, those vast frozen swamps which line the shores of the Northern Ocean, and the silvas of South America, the impenetrable forests which darken the valley of the Amazon, present examples which cannot easily be overlooked. But under the same conditions of climate the character of the soil is of extreme importance: and again, the most exuberantly fertile soil may be controlled by the character of the climate. Some geological formations, such as the new red sandstone, yield a soil which is seldom other than productive. In others, as in the coal measures or the mill-stone grits, the soil is almost always in its natural state unproductive. Others, such as the mountain limestone, produce a short, sweet herbage, and are peculiarly suited for pastoral industry. The fertile soils contain in considerable quantities the various mineral substances which are necessary for the healthy growth of plants. Sometimes the quantity of these substances, or of some of them, is insufficient. At other times, some substance, such as oxide of iron, is present in quantities so large as to be injurious to vegetable life. In other cases the soil contains, in addition to the suitable inorganic food of plants, a considerable portion of organic remains. Of this character is the black soil of Russia, the richest cultivated tract of country in the world. This immense district, which is twice as large as France, consists throughout of the richest mould; and has, though unmanured, borne for a century without perceptible injury successive crops of wheat. It contains a great quantity of organic matter, the remains of former vegetation, very minutely divided and mixed with fertilizing mineral substances. The soil is thus free and open; the air penetrates freely; and the roots of the plant have an unobstructed passage in every direction

§ 3. Even still more conspicuous than the influence of soil is the influence of climate. The vegetation of the tropics under the stimulating action of combined heat and moisture is proverbial. The frost, the killing frost, that nips the root whose fruit seems surely ripening, has furnished to the poets of northern latitudes some of their most telling similes But the power of climate is not limited to the return of labour. It affects the labour itself. The soil, including indeed climatal influences, determines the quantity of food that man's labour shall receive; but the climate, irrespective of the soil, determines the continuity of that labour.

I have already noticed the extent to which labour is both directly and indirectly dependent upon atmospheric influences, and the effect of these influences in the formation of industrial habits. There is, however, another effect of climate which, although I may somewhat anticipate future chapters, may in this place be noticed. In the more temperate climates, the cultivation of the ground and the reaping of the crop are spread over a much longer time, and can, consequently, be more cheaply effected than they can in higher latitudes. In the countries on the Danube, for example, these operations are spread over seven months; in the countries on the north of the Volga, they must be concluded in four months. Thus the work which in the former case requires four men and four teams of horses, requires in the latter case seven men and seven teams. It has been estimated* that on two estates of 600 acres each, yielding an equal gross return, and similarly circumstanced in every other respect, but situated the one at Mainz, the other in Yaroslaf, the net annual revenue of the German property would be £730 ; while that of the Russian would be £380. If it were possible to dispose of the superfluous labour and stock, it would be immaterial whether four teams were maintained for seven months, or seven teams for four months. Even in individual cases such an arrangement would seldom be easy : and for an entire district it would be obviously impossible. Many kinds of agricultural labour, too, can be carried on during winter in the southern country which cannot be attempted in the north. Even where the difference of climate is less marked than it is in the case I have mentioned, the same influence may be traced. The free-grown sugar of Honduras can undersell in New Orleans the produce of local slave labour.† Yet the cultivation of the Corosal planter is described as extremely slovenly ; his extraction of the sugar is wasteful and inartistic ; and his loss by leakage during the voyage considerable. His great compensating advantage is found in his climate, which allows the cane to ripen at any period of the year. The planter is thus enabled to distribute the work of the harvest over twelve months ; and to turn to the greatest advantage the labour at his disposal. Thus a climate in which time is not essential to industrial operations, requires, for the attainment of the same object, the use of a smaller capital than is needful

---

* Haxthausen's *Russian Empire*, vol. i. p. 150.
† Levi's *Annals of British Legislation*, vol. x. p. 389.

where the seasons within which the work must be finished are strictly defined.

§ 4. One of the chief natural advantages which any country enjoys is the presence in it of plants and animals that are suitable to human purposes. The distribution of these organic forms depends in a great measure upon conditions of climate and of soil. The influence of the former in this respect is well known. The geographical limits of the spices and of the cereals, of the vine and of the sugar cane, are as distinctly defined as those of the rein-deer or the sheep. Under the same climate, the influence of the soil on at least vegetable life is clearly perceptible. In that portion, for example, of the Southern States of America which lie between the Atlantic and the Alleghanies, the land rises in a succession of terraces, each of which has its own geological formation, its own flora and its own industry. The low but rich muddy swamps which line the shore produce spontaneously the oak and the hickory, and yield to the cultivator abundant crops of cotton and of rice. On the drier alluvial plains sugar and tobacco are successfully grown. A belt of tertiary sand succeeds, covered with pine forests, and hardly if at all invaded by the settler. These forests give place to wide treeless prairies resting upon the secondary chalk formation and producing abundantly a peculiar kind of wheat. Lastly come the primary metamorphic rocks and granite, which are naturally covered with forests of broadleaved trees, and when cleared admit of general agricultural industry.* But even where the physical conditions are alike, great diversities are observed in the distribution of plants and animals. In North and in South America respectively, there are within the latitudes corresponding to those of

---

* Johnston's *Elements of Geology and Chemistry*, p. 100.

Australia about a dozen different species of edible ruminants. In the corresponding latitudes of Africa and of India respectively, there are as many as fifty such species. In Australia,* there is not a single creature of the kind; although the country is capable of supporting more than half of the known species of ruminating animals.

§ 5. There are some natural agents whose influence in production is merely secondary. They do not themselves satisfy human wants; but they either directly or indirectly assist men in obtaining the means of such satisfaction. Of those which render such assistauce directly, the various woods and minerals, the power of water, and the properties of steam and of electricity, furnish obvious examples. Some of them are universally diffused : of others the available quantity is limited. Without the aid of these agencies, as we shall presently more clearly see, man's power over the other classes of natural agents would be very restricted. The resources therefore of a country may be regarded as more or less abundant according to the quantity that it possesses of such agents and their quality. The manufacturing prosperity of England is in a great measure due to its minerals. Its iron mines are found in immediate proximity to its coal; and abundance of lime, which serves as a flux for the ore, can be obtained from the adjacent strata. Hence the country possesses extraordinary facilities for the production of machinery, and consequently of those manufactures to which machinery has been applied. Although some portion of the coal and other minerals that are produced in England is consumed in the direct supply of wants, the largest part is used in generating or assisting to generate power, or in otherwise facilitating industry. The statistics of the British

---

* *First Annual Report of the Acclimatisation Society of Victoria*, p. 39.

Mines will therefore aid us in estimating the importance to a country of its mineral treasures. In 1854 the actual value of the minerals found in the United Kingdom, computed at the surface of the mine or in the case of pig iron at the furnace, and independently of the additional labour expended upon them before they reached the public is said to have exceeded twenty-eight and a half millions of money. This calculation does not include the clays or the building and the ornamental stones which are also of great value. The production of the ores of the metals and coals, in the actual processes of mining, and in the preparation of the ores on the surface of the mines and collieries, gave employment in that year to about 304,000 persons.*

§ 6. There are other natural agents which neither themselves satisfy wants, nor directly assist in procuring the means of their satisfaction. Their utility to man consists in the facilities which they afford for communication. In this respect also, different countries possess different advantages. During the greater part of the year, in the prairies of North America, the pampas of Buenos Ayres, or the pasture lands of Australia, vehicles can pass for hundreds of miles without any very serious obstacle. It is only when the district becomes settled, when the land on either side of the usual routes is occupied, and when consequently both the volume of traffic is augmented and its channel is narrowed, that the necessity for roads in such countries is felt. The presence also in any country of appropriate beasts of burden materially affects the power of intercourse. The horse and the ox are most in use among the natives of Europe and their descendants. The camel and the elephant have from early

---

* *Journal of Statistical Society*, vol. xix. p. 822.

times been used in the East. The rein-deer and the dog are suited to the snows of Polar latitudes. The patient and sure-footed ass or the mule is preferred for the dangerous passes of the Alps or the Andes. It is also essential that the country should yield food and drink for the traveller and his cattle. The carriage of provisions or of water increases the difficulty of locomotion: and this difficulty, if the journey be long, soon becomes insuperable.

But the great agent which nature has provided for promoting human intercourse is water. The extent of sea-coast of a country, the capacity of its harbours, the number and the navigability of its rivers and lakes exercise especially in its early history a most important influence upon its prosperity. Navigable waters have been truly* called the great highways with which Providence has pierced the desert and rendered it accessible to man. It is not until wealth has largely accumulated and knowledge increased, that art is able by elaborate and costly contrivances to compensate for the deficiency of these natural gifts. It is in those countries where the inland navigation is most complete, or where the gentleness of the ocean encouraged rather than repelled the feeble efforts and the humble appliances of early mariners, that industry has been soonest developed. It is in those countries where no such facilities exist that its progress has been most retarded. We see in America and in Australia great and flourishing cities springing into existence on the shores of the sea, or on the banks of the navigable lakes or rivers; while the less accessible districts are left to the savage and the beast. If we look back to more remote ages, we find the older civilizations clustered round the great Mediterranean lake or slowly creeping up the banks of rivers So late as the age of Augustus no place of political import-

---

* *De Tocqueville's Memoirs*, vol. i. p. 200.

ance, except some military colonies and a few towns in Asia Minor, lay at a distance of twenty miles from the shores of the great inland sea. The upper country at least in Western Europe, presented almost everywhere an interminable expanse of primeval forest, broken sometimes by wide grassy prairies, sometimes by rugged mountains, and again by impassable morasses.* In China the most populous towns are on the banks of the great rivers. The course of the rivers has for centuries directed the course of Russian colonization. A thin strip along the margin of the principal streams is fully cultivated, and the population within this district is dense; but the great and fertile plains beyond are still waste.†

§ 7. Adam Smith ‡ has noticed several conditions, the concurrence of which is necessary to render fully available any inland navigation. He points out that the rivers or other inland waters of a country ought to be mutually connected; ought to form many branches; and ought to be included in their entire course within the territories of a single nation. As an illustration of the want of sufficient connection between important rivers for a system of navigation, he refers to the great rivers of Central Asia and of Central Africa as then known. These rivers, though of great magnitude, are too far distant from each other to carry commerce and communication through the greater portion of their respective continents. Another illustration may perhaps be found in those envious necks of land that bar the communication between sea and sea; and which at Corinth, at Suez, and at Darien, have

---

* Merivale's *History of the Romans*, vol. iv. pp. 404, 422.
† Haxthausen's *Russian Empire*, vol. ii. p. 63.
‡ *Wealth of Nations*, book i. c. 3.

hitherto baffled the power of ancient conquerors, and the skill of modern engineers. The most striking instance of a great water system fully appreciated and improved is found in the great chain of lakes and the vast rivers to which North America owes so much of her prosperity. The water system of South America possesses still greater natural advantages, although it has not yet produced equally splendid results. Three mighty rivers, the Orinoco, the Amazon, and the La Plata, each navigable for thousands of miles, and each receiving the tributary waters of many noble streams, intersect the country. There needs but a canal of three miles long to connect these three great channels of communication, and to establish an uninterrupted inland navigation throughout forty-two degrees of latitude. From Demarara in the northern hemisphere, eastward to the foot of the Andes, and thence to where the Plata meets the ocean, to Buenos Ayres far below the tropic of Capricorn, the navigation would thus be complete. Russia, too, possesses in this respect great advantages. It is said,[*] though probably with exaggeration, that a communication might be opened from the Baltic to the confines of China, by means of the tributary streams and small canals. It has also been estimated, that not more than about four hundred versts (about two hundred and sixty miles) of canals would be required to enable vessels to pass from St. Petersburgh to the Pacific, a distance of eight thousand miles.[†]

The second condition of prosperous inland navigation to which Adam Smith refers is nearly related to the first. Not only is it desirable that a connection should exist between separate rivers: but each river should divide itself into

---

[*] Haxthausen's *Russian Empire*, vol. i. p. 8.
[†] Ib. vol. ii. p. 36.

numerous branches, and so form a number of navigable canals. It is to this tendency to scatter themselves in various directions through the country that Smith attributes the great and beneficial influence of the Nile and the Ganges, and still more of the Chinese rivers ; and he draws attention to the fact that "neither the ancient Egyptians, nor the Indians, nor the Chinese, encouraged foreign commerce, but seem all to have derived their great opulence from their inland navigation." In the same manner the Delta of the Niger, a space about equal to Ireland, and composed throughout of the richest alluvial soil, is said to be intersected in every direction by navigable branches of the parent stream, and thus to present an ample variety of natural channels of communication. Mention has already been made of the numerous tributaries of the great South American rivers. These rivers also form extensive deltas, and enter the ocean by many channels. In their case there seems every reasonable ground for the prediction* that "ages hence when the wilds are inhabited by civilized man, the tributaries of those three great rivers, many of which are navigable to the foot of the Andes, will by means of canals form a water system infinitely superior to any that now exists."

The third condition of a good water system relates to the political circumstances of the countries in which the waters are situated. For the purposes of inland navigation, the entire extent of that navigation should be comprised within the territory of a single Government. Were it otherwise, the communication might at any time be impeded or even wholly checked by any disagreement between the States that possess separate parts of the stream. "The navigation of the Danube," says Adam Smith, "is of very little use to the different states of Bavaria, Austria, and Hungary, in comparison of what it

---

* Somerville's *Physical Geography*, p. 263.

would be if any of them possessed the whole of its course till it falls into the Black Sea." Political considerations of a similar kind have impeded all projects for ship canals through Suez or Panama ; and may possibly hereafter interfere with the development of the water system of South America. From the same cause, in the event of hostilities between Great Britain and the United States, serious impediments to the navigation of the great lakes must arise, although by the construction of the Erie Canal the difficulty to the Americans has been to some extent obviated; and not the least calamitous event which has sprung from the disruption of the United States is the difficulty that it has occasioned regarding the navigation of the Mississippi.

The political control however of the navigable waters forms only a part of a more general condition. It is necessary that the navigation should be free from interruption of whatever kind. But interruption may be produced not only by political but by physical causes. The more free any river is from rapids, or snags, or rafts, or shallows, so much the more navigable it is: and so much the less expenditure is required for making it available for human purposes. Such obstacles, however, can generally be overcome ; and are therefore far less formidable than the influences of climate or of the character of the high lands from which the rivers flow. Many important channels of communication are for the fourth part or even for half of the year quite useless. In shallow rivers such as those in Australia and those on the Texas slope, navigation must be abandoned during the summer months. From November until May the St. Lawrence and the Erie Canal are impassable from ice ; and during the same period the same constant enemy of the seaman blockades in like manner the waters of the Baltic. During this long period these great channels of intercourse practically cease to exist.

§ 8. The subject of the physical obstructions to navigation belongs in strictness to another set of natural circumstances, different from the agencies of which I have spoken. This class comprises what may be termed the negative advantages of a country. A country may be attractive as well from the absence of obstacles as from the abundance of its powers. There are some countries which are sterile from the superabundance of water; there are others which are parched from excessive drought: in the one case time and trouble and labour must be expended to effect a system of drainage or of embankments; in the other case costly works of irrigation are required. In some countries dense forests, through which a path must be cleared with steel or with fire, impede the labourer. Sometimes the intense heat or the intense cold prevent, during a considerable portion of the year, all industrial exertion. There may be, indeed in connection with some of the preceding cases there always are, terrible diseases: ophthalmia, malaria, fever, ague, dysentery, and a thousand forms of death. There may be the plague of insects; there may be venomous reptiles, and noxious beasts, and, most formidable of all, untamed and ferocious men. Or again, some great elemental convulsion may be of frequent occurrence. There may be sudden and destructive floods; there may be hurricanes; there may be earthquakes. The presence of any of these phenomena is, so far as it goes, a deduction from the natural advantages of the country. If it can be removed, the removal involves considerable expense. If the evil be incapable of remedy, it often exacts a terrible tribute: and even beyond its actual pernicious effects, the apprehension of uncertain and irrepressible ill tends to repress and discourage industry. In this case, although in a somewhat altered sense, the rule *magnum vectigal est parsimonia* is true. In those favoured countries where natural obstacles do not exist, or if they do exist are found in a very modified form, the community has at its disposal a great fund for other industrial purposes. The war against the wilderness that every new settlement must carry on, is then waged at the lowest cost and at the greatest advantage. The money, the time, the strength, the lives, that would in other circumstances have been expended in draining or in irrigating the fields, in clearing the forest, in extirpating ferocious beasts, or in warring with the savages ; or which would have been consumed by disease, or wasted by compulsory idleness, or overwhelmed by the fury of the elements, or cut short by sickness and untimely death, are available for a better culture, and for the exploration of new fields of enterprise.

§ 9. As water transit served to illustrate the natural advantages which relate to communication, so an illustration of negative advantages may be found in the principal conditions of land transit. These conditions may be stated indeed in a positive form, but they are best known to us as negations. They consist rather in the absence of obstacles than in any direct advantage. The surface of the country over which commodities must pass, should for the purposes of traffic be level, clear, smooth, and hard. In other words, it should neither be secluded from other countries by mountains, nor incumbered with forests or with scrub. Nor should it present tracts of rocky or broken ground ; nor morasses ; nor sand ; nor mud ; nor rivers which cannot be forded ; nor lakes round which a circuit must be made. Of these impediments to transit it will suffice to mention two or three instances in which they have produced important consequences.

The difficulty of communication between its different regions is one of those characteristic features of Eastern Asia which have largely influenced its history

One road only, that through the Valley of Peshawar, leads from Persia to India. There is no practicable road for regular commerce between India and China; and the communication between these great peninsulas has always been by sea. The passage of the Great Plateau is always a difficult and tedious undertaking; and at some points is almost impossible. Thus Eastern Asia has always been isolated ; and the development of its several countries has consequently been imperfect. In this case, as indeed in almost all other cases of the same kind, the political results of this difficulty of communication greatly transcend any mere economic consideration. We have another example in a country whose history is to us better known, and not less interesting than that of the monotonous East. In the Peloponnesus, irregular mountains, valleys frequent but isolated, land-locked basins, and declivities which often occur but seldom last long, form the general character of the country. Its disposition and properties do not admit of permanent rivers, but are favourable to the formation of lakes and marshes. Hence there was little motive and still less of convenient means for internal communication among its inhabitants. "Each village or township occupying its plain with the enclosing mountains supplied its own main wants, whilst the transport of commodities by land was sufficiently difficult to discourage greatly any regular communication with its neighbours. In so far as the face of the interior country was concerned, it seemed as if nature had been disposed from the beginning to keep the population of Greece socially and politically disunited, by providing so many hedges of separation and so many boundaries, generally hard, sometimes impossible to overleap."*

Instances of the obstacles to communication that arise

---

* Grote's *History of Greece*, vol. ii. p. 294.

from tangled underwood or dense forest abound in most new countries. In such cases, the traveller is not unfrequently compelled at every step to clear his way with the axe. But the force of the obstacle is more keenly felt when it arises in a country in other respects tolerably free from its effects. In Canada, the question of the clergy reserves was for many years a subject of bitter dispute. Extensive tracts of land had been granted for ecclesiastical endowments, and in most instances continued in their wild state. The Imperial Government and the Church were reluctant to alienate these lands, which at some future time might become very valuable, but which the clergy could not immediately use: the colonists were urgent in desiring that the reserves should be rendered productive. The real grievance in this matter consisted, not in any offence occasioned by the endowment or its distribution ; nor in the withdrawal from production of so much land ; but in the fact that these uncleared reserves were impenetrable barriers between adjacent settlements. In addition to the impediments that the forests presented to travelling, they also afforded a shelter to wolves and other dangerous animals. Thus, both by the actual distance and by the intervening forests, and by the wild beasts that found shelter in those forests, free communication between the settlers was seriously impeded.

Another important condition of facile communication is, that the country should be free from such noxious animals or vegetables as are likely to impede the traveller. Against the larger and more formidable animals it is generally possible to take sufficient precaution ; but the inferior forms both of animal and of vegetable life sometimes present almost hopeless obstacles. To the former class belongs the tsetse or poison fly, the terror of the South African wilderness. This terrible pest, which haunts extensive districts of South Africa, is about the size of the common blow-fly; but has a poison equal to that of the most venomous reptiles. Its sting, although not injurious to man or even to the wild animals, is fatal to horses, dogs, and oxen. African travellers tell how adventurous settlers are compelled to return in haste from the journeys they have commenced because they have met with the country of the tsetse. Dr. Livingstone, although he used the utmost vigilance, and although not twenty of these pests alighted on his cattle, lost in a short time forty-three oxen. He also mentions that on one occasion a native chief lost from the same cause nearly all the cattle of his tribe; and there are large tribes which are obliged to abandon altogether the attempt to keep cattle or sheep. Of the pests that belong to the vegetable kingdom, parts of Australia afford examples. In the warmer latitudes of that country a plant *(strychnus lucida)* is found, which contains strychnine in considerable quantities, and has often caused great mortality among flocks and herds. At Port Essington it is so common that cattle, to ensure their safety, must be kept in enclosures, and supplied with artificial food. In some districts on the Upper Murray and the Darling, there are plants which produce in horses strange and fatal forms of madness. In parts of West Australia, little pea-like flowering plants, *(graptololium bilobum)* containing a deadly poison, are found in such abundance as to render the districts in which they occur useless for pastoral purposes. Even in travelling through these districts carriers have lost in a single night, from the effects of the poison plants, three-fourths of their team. In other parts of the same country where pastoral pursuits have been hitherto safely carried on, cases of apparent poisoning have occurred. It has been supposed that in these localities poisonous fungi parasitic on fodder plants have become developed; but no accurate knowledge on this subject appears to have been as yet obtained.

§ 10. Such are the principal circumstances that affect the absolute capabilities of natural agents. But their efficiency, that is, their utility to man, involves other considerations. The most potent natural agents are not always the most available. The terms productive and profitable are not convertible. It is manifest that no degree of excellence in a particular natural agent can compensate for the absence in its proprietor of any desire at that time for that agent. The best wheat would be useless to a wanderer perishing from thirst. The purest gold coin serves only to mock the wretchedness of the ship-wrecked seaman. But even where the agent is of a suitable kind, proof of its excellence in its kind is not conclusive proof of its fitness. There is no country in which all the natural agents it contains are fully used. There is none in which all those that are so used are the most potent that the country contains. It has often been noticed with surprise that countries with great natural advantages do not become the seats of those manufactures for which they seem to be peculiarly adapted; and many unsuccessful enterprises have been undertaken on account only of the abundance in the locality of the natural agent chiefly required in the business. But experience has amply shown that such an advantage, however important it is, may not be alone sufficient. The greatest water-power in the world is scarcely if at all applied to industrial purposes. The English potteries are not located in the counties which produce their clays. The cotton trade has not taken root in Georgia or South Carolina. There are no woollen factories in Australia. The fisheries of Norway and Ireland are worked not by local enterprise, but by English capital for the London market. So it is with land. There is perhaps in the whole world no land equal in fertility to the Silvas of the Amazon. They occupy a country more than six times the size of France, under a tropical climate, with a soil of the richest mould, abundantly watered by periodical rains, and intersected by the largest river in the world and by its numerous tributaries, the least of them a noble stream. Yet this territory of unequalled facilities remains one continuous expanse of primeval forest: and its inhabitants are still the jaguar and the ape.

One cause of this apparent neglect or apathy is found in the possible co-existence of the different classes of natural agencies. The physical advantages of a country depend, not upon the presence of any one of these classes, but upon the concurrence of them all. That country is most useful to man which enjoys the greatest number of the advantages that are included in each of these different classes: that is to say, which has the largest amount both of those agents that are directly concerned in production, and of those that are auxiliary to it; which possesses the greatest facilities for communication: and in which no deduction is made from the sum of these advantages in consequence of the presence of hostile or obstructive influences. Some important consequences of this principle in relation to manufacturing industry will be considered in a subsequent chapter. The vital importance of situation, and the extent to which this advantage outweighs extent of territory or fertility of soil, have been already noticed. But even where no question of situation arises, the superior agent may be injurious to health; or may be expensive to work; or before it can be at all brought into operation may require a large preliminary outlay. Humboldt* lays down the rule that in tropical countries great fertility of soil and unhealthiness of atmosphere are inseparable from each other. In almost every country extensive tracts of land on certain geological formations are laid down in natural grass on account of the difficulty and

---

* *Aspects of Nature*, vol. 1. p. 33.

expense of working them. Such are the Lias; and the Oxford, Weald, Kimmeridge, and London Clays.* In the chalk formation of Alabama water can only be obtained by deep sinking. This process is very costly, and thus the State is suited only for wealthy settlers.

The distinction between the potential and the actual utility of land is forcibly expressed in the following remarks of Professor Johnston† upon a district in New Brunswick: "I may advert here to a reflection which often crossed my mind as I travelled over this and other parts of the newer countries of North America that an important distinction must often be drawn between the actual or present and the future or possible capabilities of tracts of land which lie on the same geological formation, and of which the soils possess the same chemical and mechanical characters. Absolutely considered, soils which have the same geological, chemical and mechanical relations, ought to be equally productive. But if their natural conditions be unlike, in respect for example to the drainage of water, one may be of great immediate value, and be with little time and little cost rendered capable of supporting a large population; the other may be wholly useless, and may lie barren and unimproved for numerous centuries. Thus much of the absolutely good and capable red land of the New Bandon district in the Bay de Chaleur, and still more perhaps of the heavier land between the Napan river, and that of the Bay du Vin is too wet for cultivation and often covered with swamps because it is too low to allow the surface water a ready means of escape. Yet this swampy and inhospitable tract if laid dry is as susceptible of agricultural improvement and of being made a source of rural wealth as the apparently

---

* *Johnston's Elements of Geology and Chemistry*, p. 114.
† *Notes on North America*, vol. ii. p. 48

richer patches which rise above the common level and naturally free themselves from superabundant water. Contrary therefore to the indications of both geology and chemistry, thousands of acres in these countries which will at some future period yield abundant crops, must in actual circumstances be pronounced to be almost worthless."

§ 11. It must not be forgotten that most of these natural facilities and impediments are relative. Their influence whether for good or evil depends in a great degree upon the wealth and still more upon the knowledge of the people among whom they exist. It has been observed that while the gentler inland seas have encouraged early navigation, the ocean long presented an insuperable barrier to its progress. Although the real difficulties and dangers of oceanic navigation probably exceeded the limited means which most nations of antiquity possessed, yet these obstacles were trifling in comparison with the imaginary terrors that appalled the stoutest hearts among the mariners of Hellas or of Rome. The fog, the mud, the shallows, the seaweed, the thousand nameless terrors that almost drove into mutiny the crew of Columbus, sent back even to a certain and a cruel death the less fortunate Persian* adventurer who was condemned, as the severest penalty that could be devised for a heinous crime, to explore the mysterious Indian Ocean. There seems indeed to be good reason† to believe that two thousand years before the great achievement of De Gama, the Phenician mariners in the service of the King of Egypt had rounded from the East the Cape of Good Hope. But while the discovery of the Portuguese Admiral affected the destinies of the world, the same discovery of the officers of

---

* Herodotus, iv. c. 43.
† Grote's *History of Greece*, vol. iii p 382.

Nekos remained as barren as the discoveries of M'Clure and M'Clintock in our own days must remain. In each case conclusive proof was obtained as well of the possibility of the feat as of its inutility. There was no physical impossibility to the Egyptian king in the circumnavigation of Africa, as there is none to us in the north-west passage; but the undertaking cost more than it was worth. The passage from the Mediterranean to the Red Sea by the Nile and overland was with the resources that these early mariners possessed incomparably easier, and therefore better adapted to commercial purposes, than the ocean route with which we with larger vessels and better appliances and greater knowledge, although not with greater courage, than the gallant Phenicians are now so familiar.

Even the navigation of the Mediterranean was in ancient times dangerous and slow. In the seventh century before the Christian era, nothing short of Divine command through the Delphian Oracle, in addition to the sufferings of a seven years drought, could have persuaded the people of Thera to attempt the almost unheard-of feat of crossing the open sea to Libya. Six hundred years afterwards, the passage from Cadiz to Rome took three months in certain seasons of the year. Yet when the properties of the magnet became known, and the arts of naval architecture and of navigation were improved, that ocean which had once seemed the divinely appointed limit to human effort, and a voyage upon which Xerxes and his Court regarded as a punishment worse than death, became the great highway of the nations. In like manner the river or the lake which in more favourable circumstances is the greatest natural advantage that a country can enjoy, checks the advance of the wandering savage or the isolated traveller. It is the same with other impediments. There are few material obstructions which the combined influence of engineering skill and wealth cannot now remove. The rivers can be cleared of snags or of bars; and the rapids evaded by canals. Arms of the sea can be spanned by bridges at such an elevation that ships can sail below. The forests can be cut down; the hills can be avoided, or levelled, or pierced through, or surmounted. Secure foundations can be laid and a smooth hard level surface obtained in the morass. Wells are sunk and hotels established in the desert; beasts of burden if they are not indigenous can be imported; and noxious animals and deadly vapours recede as the forest is cleared or the marsh drained.

CHAPTER VI.

OF THE MEANS BY WHICH THE EFFICIENCY OF NATURAL AGENTS IS INCREASED.

§ 1. Since the intrinsic efficiency of natural agents depends upon their quantity and their quality, the modes of increasing this efficiency must relate either to the augmentation of the one or the improvement of the other. Of the advantages consequent upon the increased quantity of natural agents directly or indirectly engaged in the work of production, the examples are obvious. There may be in the first instance the discovery of new countries, of portions of the earth previously unknown or unavailable for industry. Such on a great scale has been the discovery of America, of Australia, and of New Zealand. Such too has been the discovery, in parts of each of these countries beyond the limits of previous settlement, of fertile and convenient localities. And such has been the discovery, for to Europeans it is practically a discovery, of the populous inland regions of China and of Japan. Even the discoveries, at first sight so unpromising, of the intrepid voyagers in the Arctic Seas have opened new fields for the cramped energies of the hunters of the whale. To the over-crowded pastoral tenants of New South Wales and Tasmania, the discovery of South Australia and of Victoria gave control over an increased quantity of the natural agents that they desired; and thus increased their productive power. When the pressure of population began to be felt in these new colonies, Queensland and the recently discovered fertile lands in Central and Northern Australia, so long regarded as desert, have rendered to them a service similar to that which they in former days rendered to the older settlement. So too in America the great statesmen of the revolution knew nothing, when they declared their independence, even of the sites on which Cincinnati and Chicago now stand. From the commencement of the present century to the time of their fatal quarrel, the United States settled with a thriving population a territory exceeding 360 millions of acres, the greater part of which was a hundred years ago absolutely unknown. As far as the production of wealth is concerned, the abundant natural agents that these territories contain had previous to their settlement practically no existence. The scanty tribes of wandering savages that have generally inhabited newly discovered regions, have been unequal to use the riches that surrounded them; or where the original inhabitants have made a greater advance towards civilization, their knowledge of industrial arts has been imperfect. Thus in all such cases of discovery, either the more intelligent race has supplanted the inferior race in the possession of those natural forces that the latter could not use; or a stimulus has been given to the more civilized natives sufficient to rouse their feeble industry, and to promote their material advancement. But whether the resources of the newly discovered country are developed by its discoverers, or by the new-sprung activity and enterprise of its former inhabitants, the general gain is alike secured. That which formerly was unused has become productive. No statist would venture to estimate the addition that has thus been made to the wealth of the world.

Sometimes the discovery of new countries brings with it the discovery of new plants and animals. Thus it is to the discovery of America that we owe the discovery of the potato, the tobacco, and the maize; the alpaca, the beaver, and the sable. But we often discover in well-known plants properties of which we were previously ignorant; and we still more frequently enrich our own country with the indigenous plants of countries long known and settled. The cereals and the fruits of Britain, themselves often derived from other lands, are flourishing at its antipodes. The cotton plant and the sugar cane of the East are now the staple products of the American Slave States and of the West Indian Islands. A few years since a surgeon at Singapore happened to observe the peculiar handle to the axe of a Cingalese woodcutter; and from this casual observation a great trade in gutta percha has sprung into existence. An escaped slave brought from Asia Minor to the heedless court of Louis Quinze a few madder seeds; and before he died, not however without many repulses and often many losses, saw a new branch of industry enriching France though not himself. In like manner the useful animals of other countries have been introduced into lands far remote from their original seats. Many parts of the earth, though well suited to certain kinds of animals, are not inhabited by these kinds. Some countries which have similar conditions of climate and of surface, are separated from each other by barriers insuperable to the lower animals. Such countries contain not the same but cognate animals and plants. But since the conditions of existence in all these cases are substantially alike, it becomes possible to transfer from their respective abodes into any one such country all the representative species, as they are termed, of other countries. Various countries accordingly have been and now are enriched by the introduction of suitable animals. A missionary at great personal

H

hazard brought in a cane to Europe from China the eggs of the silk worm: and from this small beginning the present gigantic trade in silk has been developed. Mr. McArthur brought the merino into Australia; and so gave rise to the great trade in wool. The camel has been brought from his native seat to America and to Victoria. The horse has multiplied in countless numbers on the Pampas; and many tenants of British air and British waters, and many valued animals unknown to Britain, are gradually finding a new home in Australia.

But if we now rarely discover new plants, and still more rarely new animals, we have no cause to complain of our infrequent discoveries of new forces, or of new substances. The continual additions to our knowledge of astronomy, of steam, of magnetism, of electricity, of the properties of gases, of the chemical properties of the soil, of the more intimate structure of the vegetable and the animal organism; have all extended to an astonishing degree the dominion of man over nature; and have rendered available new means for the supply of human wants. New substances make their appearance from the most unexpected quarters. The accidental combustion of sea-weed on the sandy beach gave rise to the use of glass. The application of snow-water to a bronchocele led to the discovery of iodine. Some chemical experiments resulted in the use of chloroform. The whole course of the discoveries in physical science has tended and still continues to tend either directly or indirectly to increase the great sum total of human enjoyments.

Another species of discovery relates to that class of natural agents, the use of which chiefly consists in facilitating communication. Of this kind the principal instances in former times have been the discovery of the polarity of the needle, and of the passage to

India by the Cape of Good Hope. Perhaps we should add those discoveries of the Alexandrian astronomers, which, though not so intended by them, ultimately rendered possible the art of navigation. In our own times we have the discovery of the electric telegraph, and of the uses of steam to the purposes of locomotion. We have also many important explorations of various lines of route, such as the Overland Route; and the discovery of navigable rivers in Central Africa. Such too are Commander Maury's discoveries in the Physical Geography of the Ocean, which have reduced by many days the time of maritime passages. On a smaller scale, but often of greater immediate importance, are all explorations to ascertain the navigable capacity of a river, or a better line of road, or a convenient pass over mountains or through a swamp. The removal of bars and snags and other obstructions in rivers, the cutting of canals to avoid cataracts and rapids, the erection of piers and light-houses, the projects so long and hitherto so unsuccessfully entertained of cutting through those narrow barriers that prevent the insularity of Africa and of South America, may all be taken as examples of increased efficiency being given or attempted to be given to natural agents for the purpose of communication.

§ 2. Great as has been the gain to production from discovery, still greater gain has been effected by improvement. By this term I mean not the discovery of natural agents previously unknown or unused; but the knowledge of new combinations, or of new applications, of agents already known. As the former class increases the quantity of natural agents available for man's purposes, so the latter class improves their quality. The improved agent thereby becomes more serviceable to man than in its original condition it had been. Those improvements which increase the efficiency of the actual agent are, in the sense in which the terms are here used, distinct from those inventions the utility of which consists in the abridgement of human labour, and the substitution for it of physical forces. There is a good illustration both of discovery and of improvement in the history of India rubber. This useful substance was not discovered until a comparatively recent date. Its properties were little known; the purposes to which it was applied were few; and its consumption was consequently limited. In 1830 the quantity imported amounted to about 50,000 lbs. About that time an improvement in its preparation was effected by the method known as vulcanizing the material. This improvement greatly increased its utility: but a few years afterwards a further improvement was adopted. The vulcanized India-rubber was prepared with sulphur; and the material thus improved became applicable to a multitude of purposes. Its consumption has risen from fifty thousand pounds weight to two millions of pounds; and the trade is so important as to have obtained a special history of its own.

Perhaps no better example of improvement can be found than irrigation. In that vast belt of desert which extends from the shores of the Atlantic to Arabia, where no river flows and where no rain falls, the soil seems condemned to perpetual sterility: and yet wherever the ground can be irrigated by wells, the sands can soon be made productive. In India some of the greatest works not only of ancient but of modern times are constructed for the purposes of irrigation. It has been estimated that in the great famine of 1837-38, the gross value of crops grown on land irrigated from the united Jumna canals, the greater part of which land would otherwise have been totally unproductive amounted to about two millions sterling.* The anicut on

---

* *Journal of Statistical Society*, vol. xxi. p. 180.

the Godavery has rendered productive a million and a quarter of acres that were formerly barren. An area of not less than five and a half millions of acres has been rescued from sterility by the stupendous Ganges canal; and a check has thus been given to the periodical famines in which millions of lives and vast sums of money have been lost.

But if water be deficient in some places, in others it is found in excessive abundance. This superfluity must be removed before free scope can be given to the productive powers of nature. Sometimes the superfluous moisture is found beneath the ground; sometimes it comes either constantly or at periodical intervals to the surface: in some cases the cause of the inundation is a lake or river: in other cases the sea itself is our opponent. It is needless to describe how the Low Countries have to a great extent been recovered, and by unceasing care preserved, from the inroads of the German Ocean. Within the last thirty years upwards of two millions of acres have, by means of surface and thorough drainage, been for all productive purposes added to the size of Ireland. On the land which is now the very garden of England there still stands the lighthouse* which only a century ago was erected to guide the wayfarer through swamps as trackless and as dangerous as the sea. The impassable morass which at the battle of Preston Pans protected the front of the Royal army against the charge of the Highlanders, and across which a friendly resident, by a path known only to himself, guided Charles Edward, now forms part of one of the most fertile farms of the Lothians.†

Another remarkable instance of improvement occurs in the use of the hot-blast. It was found that in smelting

---

* *Journal of Agricultural Society*, vol. iv. p. 287.
† See Smiles's *Lives of the Engineers*, vol. ii. p. 108.

iron the power of the furnace was greatly increased by heating, prior to its admission, the air with which the furnace was supplied. Careful experiment has shown that where, in the production of one ton of pig-iron, three tons of coal with certain quantities of calcined ore and limestone were consumed, a similar quantity of iron can be produced from the same quantities of ore and limestone, under the influence of the hot-blast, by the consumption of only one ton and three-quarters of coal. But besides the great saving in the consumption of coal, the introduction of the hot-blast produced other and still more important results. Under the old system of smelting, anthracite was too stubborn a kind of coal for convenient use as a fuel; and black band iron-stone was too stubborn a kind of ore to be freely used by the smelters. Since the hot-blast has been introduced, the anthracite yields an intense heat; and the black band produces a rich per centage of good iron. It is said that the market value of the anthracite of South Wales and that of the black band of South Scotland have been increased by millions sterling in consequence of this great improvement.

§ 3. The efficiency of natural agents, or in other words their capability of satisfying human wants or accomplishing human purposes, does not, as we have seen, depend exclusively upon the quantity or the quality of these agents. Such efficiency is not absolute but relative. There is a distinct relation between the natural agents and the wants and purposes which they are to subserve. It is not enough that there should be natural agents capable in certain circumstances of ministering to human requirements; but these agents must be capable of ministering to such human wants as men at the moment feel, or such purposes as they are at that time or that place prepared to execute. Nature yields her assistance only on the fulfilment of certain conditions; and these conditions are onerous according to the magnitude of the aid. A natural agent is useful to man in proportion to its force: the greater the force, the greater will be the benefit when it is controlled and directed to human purposes; but at the same time the greater will be the difficulty in obtaining over it the needful control. But man when left to himself is the feeblest of all the larger animals. The powers of all other living creatures are proportioned to the circumstances in which they are placed: man alone is unequal to the continued struggle with nature. He either quickly sinks in the unequal contest; or with difficulty and painfully protracts a miserable existence. In such cases the natural forces which he uses are those supplied by the spontaneous bounty of nature. When he first advances beyond this condition, the forces which are available to him are of the very simplest and least efficient kind. Such as they are however, they assist him, either directly or by setting free a portion of his labour, in obtaining other and more powerful auxiliaries; and these in turn form another step in the slow and difficult ascent to wealth. The natural forces at his disposal become continually more efficient than those which he had previously been able to use, because he is able from time to time to fulfil with the appliances at his command the conditions upon which these superior forces may be rendered available. The natural order therefore of man's dealings with physical forces is from the less powerful to the more powerful; and his advance is both gradual and sure. It is gradual, because his strength is unequal to reach from one state to another remote from it without previously passing through the variety of intermediate states; and it is sure, because each step in advance that he makes renders more easy each succeeding progression. In this case as in every other, the first step is the most difficult. The beginning, as the old philosopher used to say, is the half of the whole.

§ 4. There can be little need for any extended illustration of these principles. Subject to that great exception which I shall presently notice, few will be found to dispute their general application. When we say that civilized man is better fed, better clothed, and better housed than a savage, we mean that his wants either actual or potential in these respects are more abundantly, more regularly, and more completely supplied. If we ask how this result is obtained, the answer will be, that as man recedes from a state of barbarism, his power over nature increases : the forces at his disposal are more potent than before ; or his control over them is more complete. The savage feeds upon wild fruits or game ; his shelter is a hollow tree or at most a wigwam of sticks ; and his clothing, if any, consists of undressed skins, or of plaited grass, or the leaves of some broad-leaved tree. Very different are the resources of modern civilization. The fruits and animals which supplied the savage's scanty support are by careful tending so improved both in quantity and in kind as to be hardly recognizable. Other and more nutritious kinds of food are acquired. In procuring them advantage is taken of the most powerful natural forces. Their preparation calls forth no small degree of ingenuity and skill. Still more marked is the change in clothing and in shelter. New agents are found for materials : new agents assist in the preparation of those materials : new combinations and forms of beauty are discovered for their composition. To take a single but memorable instance, we see how in the cotton manufacture, that great type of British industry, the motive force was at first manual labour, then the force of horses, afterwards water, and lastly steam. Nor is it improbable that electricity may at some not distant day be brought to render its aid in this great manufacture. So it is with illumination. The splint of resinous wood, the oil, the tallow, the candle, the lamps in all their variety, the gas, and, to crown all, the electric light, furnish a striking example of humble beginnings gradually developed into magnificent results. Those difficulties which beset every new enterprise, to use the words of Burke, "it has been the glory of the great masters in all the arts to confront and to overcome ; and when they had overcome the first difficulty, to turn it into the instrument for new conquests over new difficulties : thus, to enable them to extend the empire of their science, and even to push forward, beyond the reach of their original thoughts, the landmarks of the human understanding itself." *

§ 5. It is possible perhaps to trace the moral influence which this sequence exercises upon man. The advance from the inferior to the superior agent is gradual, even slow ; but it is sure. In ordinary circumstances, if men do not wholly yield to the promptings of appetite against the dictates of reason, their wealth accumulates, and consequently their means of obtaining a command over nature increases. But the slowness of this process, certain though it be, arises from the pain and difficulty of the effort to practise forbearance and self-control. These qualities are essential not only to the accumulation of wealth which supplies the sinews of the further war against nature, but also to the patient and persevering pursuit of knowledge by which and which only the victory can be won. It is to the formation of these habits that our external circumstances are made to render such assistance. Men have in their self-interest the

---

* *Burke's Works*, vol. iv. p. 288.

strongest incentive to acts of self-control. The necessity is manifest; and the reward is speedy and sure. Every person, however illiterate, quickly comprehends and vividly realizes these truths. But if the superior agents were first brought into use, there would be little occasion for the exercise of the mental and moral faculties. When these agents were sufficient for men's wants, there would be no need of further exertion. When they began to fail, and labour as it increased became constantly more and more unsuccessful, the stimulus of hope and of success which men now feel would be removed; and the fear of want would be the sole spring of action. The utmost that in such circumstances we could expect would be that the species should not retrograde. If it were able to maintain its former position, no inconsiderable success would have been achieved. But by proceeding from the less to the greater, a guarantee for the progressive improvement of the race is given: and at the same time the present difficulties of his position, and the prospect of being able to diminish them, continually call forth the latent powers of each individual. It is not in the most favoured regions of the earth that the higher qualities of our nature have found the kindliest soil. Nations are but assemblages of men; and men have their powers most developed when they are most required to use them. "Difficulty," says Burke,* "is a severe instructor set over us by the supreme ordinance of a parental guardian and legislator, who knows us better than we know ourselves, as he loves us better too. *Pater ipse colendi haud facilem esse viam voluit.* He that wrestles with us strengthens our nerves and sharpens our skill. Our antagonist is our helper. This amicable contest with difficulty obliges us to an intimate acquaintance with our object, and compels us to

---

* Vol. iv. p. 288.

consider it in all its relations. It will not suffer us to be superficial."

§ 6. One large and important exception is usually made to the rule that man proceeds from forces of an inferior to those of a superior potency. This exception is land. From the time of the French Economists up to the present day, the most distinguished writers on Political Economy have generally held that land is governed by distinct and peculiar principles. It is said that men always cultivate in the first instance the best soils; that for the use of the appropriated natural agents they pay to the owner a certain sum under the name of rent; that the cause of rent being paid is the necessity for having recourse to inferior soils; that the degree of inferiority is the measure of the rent paid for the use of the better soil; and that this necessity arises from the peculiar condition of agricultural production, namely, that in a given state of agricultural skill every increase of labour upon the same land yields a product, increased indeed, but not increased proportionately to the increase of labour. According to this view, the tendency of agricultural produce is constantly to diminish: and this tendency is counteracted by those various agencies which taken collectively constitute as it is said the "progress of civilization." This principle of diminishing returns Mr. Mill declares to be the most important proposition in Political Economy. But notwithstanding the eminent names by which this theory has been supported, and the important consequences to which a departure from it is supposed to lead, a question has been lately raised by American and French writers whether the case of land is really exceptional; whether the order is from the more fertile to the less fertile, or is, as with other natural agents, from the less efficient to the more efficient; and whether the law of diminishing returns is true, at least so far as it has been supposed to be distinctive of agriculture.

Where a man has his choice of more fertile and of less fertile land, if he have the power to cultivate either of the two, and if such cultivation in other respects be in each case equally advantageous, he will undoubtedly select the more fertile. The term, "more fertile land" implies that the land with equal labour is capable of returning a larger amount of produce. But this proposition assumes not only an equality of skill but also an equality of power. A man will choose the land not which is absolutely the most fertile, but which, after a due estimate of the other classes of natural agents, is so relatively to his then existing means. But as his powers are continually expanding, his ability to take up land of which the absolute fertility is greater than that which he formerly possessed also increases. The most fertile land is seldom fit for immediate cultivation. It is sometimes difficult to work. Frequently it requires a large preliminary outlay. It is often unhealthy. The person therefore, who desires to cultivate the most fertile land must generally submit to considerable delay, and a heavy expense in clearing or draining the land or otherwise bringing it under cultivation. On the other hand the person who seeks a speedy crop at the least possible outlay must be content with a light and poor soil. Hence a young and therefore poor community must content itself with the quick though scanty returns of comparatively barren land; while the rich lands remain unused until in the course of years the wealth of the people and their aids to labour gradually accumulate. Mr. Ricardo and his followers in effect recognize these principles when they admit that land may be inferior not only in point of fertility but in point of situation. Situation is only one, although a very important one, of several conditions. If natural fertility be counterbalanced by the difficulty of procuring the assistance of exchange, it may also be affected by the difficulty of procuring the assistance of capital or of invention, or of co-operation. The negative advantages of land are not less influential than its positive facilities for communication. When a man is about to settle upon land, he has to consider its fitness for his purpose. This fitness implies not only its natural capacity, and its conditions both positive and negative, but its relation to the appliances at his command. As these appliances are extended, as the capital of the landowner is augmented, as his tools and machinery improve, as he can rely with complete confidence upon the co-operation of other persons, and as the facilities for exchanging the produce of the land increase, the difficulties that formerly obstructed his use of the more fertile land gradually disappear; and the natural advantages of such land come into full operation. In this case therefore, as in every other, and for the same reason, the natural order is from the less to the greater. These considerations explain the occurrence of what might seem at first sight an exception, but which is in reality a further confirmation of the principle. When a new country is settled by civilized men, if there be available, as in such circumstances there usually is, a considerable amount of capital and skill and associative power, the natural fertility of the lands first occupied will be much greater than it is in a country in the early occupation of which labour is not so aided. Agricultural improvements then are not merely the counteracting agencies to a constantly diminishing return; nor is the course of such improvements in any respect exceptional. The tendency of agricultural labour is to obtain from the same land larger returns, because the assistance which the latter renders is continually increasing; and because by means of such assistance he is enabled constantly to cultivate lands which notwithstanding their great potential fertility had previously, in the absence of efficient auxiliaries, been beyond the reach of at least profitable cultivation.

§ 7. These reasonings require of course the support of actual experience. It is upon experience that Mr. Carey, who first stated, although not perhaps with all the precision that could have been desired, the principles now under consideration, mainly rests his argument. That gentleman has traced the course of settlement in almost all known countries;* and as the result of his inquiry asserts that it is an historical fact that the order of settlement is invariably from the less fertile to the more fertile soils. He challenges his opponents to produce a single exception to this rule, a challenge which so far as I am aware has not hitherto been accepted. He further asserts that as the wealth of a community increases, men gradually leave the inferior soils which they had originally occupied ; and that on the contrary when from any cause their wealth is diminished, their tendency is to abandon the richer lands that they had cultivated and to fall back upon the poorer soils of an earlier generation. So it is in India ; so it is in Italy ; so it is in those wretched Eastern countries where once flourished the mightiest monarchies of the Ancient World.

Upon those accounts of the early stages of settlement which are derived from remote ages or from countries that are still comparatively unknown, little reliance can be placed. The evidence however of the countries which are best known to us is very strong. Throughout the whole West of America, both at the South and at the North, the pioneers seem invariably to choose for the commencement of their operations the inferior soils. The oldest settlements are always found upon the thinly wooded and com-

---

* *Social Science*, vol. i. pp. 107–146.

paratively barren hill lands, or upon the dry and upland prairies. It was the sandy plains and pine barriers of Georgia, of Alabama, of Florida, and of Mississippi, that received the first emigrants. The first homes of Texas were built on the upland prairies; the first homes on the Mississippi River were placed on the high clays and the rocky bluffs which now form the poorest soils in the vicinity. In Arkansas and Missouri the first settlers are found among the pine lands and the hills. The first town sites in Ohio are on the poorest agricultural regions on the river; and the deserted homes of its first cultivators are still seen upon the rugged hill sides: while cultivation has abandoned them for the rich valley lands which to the pioneer were not worth the surveyor's fees.

In Australia the natural course of settlement has been disturbed by the action of the Government in selling or withholding from sale at its discretion waste lands. In Victoria, where this disturbing force has been more acutely felt than in the other colonies, the new markets which the various gold-fields supplied further determined the direction which agricultural industry should take. But although the evidence from Australia is on this account less convincing than it otherwise would be, yet subject to these disturbing influences the operation of this principle may be traced. The lands which in Victoria were first cultivated were the poor soils on the tertiary formation south of Melbourne, and the shallow soils on the treeless lava plains to the north. But the rich lands are for the most part covered with gigantic Eucalypti, and with an undergrowth of Sassafras, Tree Ferns, and other dense vegetation. Such lands occur on the slopes of the Dandenong Ranges, on the spurs that run from the mountains of the Upper Yarra, on the ranges and the rivers of Gipps Land, and on the coast ranges in the vicinity of Ballaarat, northward and eastward of Mount Warrenheip. But notwithstanding their exuberant fertility, all these lands, except some portions of the last named locality which have very recently been brought within the influence of the railway system, remain in their primitive condition.

British North America also confirms this rule. After describing a certain kind of land, Professor Johnston* observes: "This land is now valued by the holders at 10s. an acre. It will take £2 an acre to clear it; but it will grow all the crops suited to the climate, and it gives a first crop that usually pays the whole expense. In a new country and among poor settlers this is called good land. Poor land among them is a relative term. Land is called poor which is not suitable to a poor man; which on mere clearing and burning will not yield good first crops, and which requires to be stumped and ploughed before profitable crops can be raised. Larch and hemlock land are often of this kind. The thin upper soil on which these trees grow is not rendered fertile by mere burning the wood upon it; a new soil must be turned up first. Thus that which is poor land for a poor man, may prove rich land to a rich man who has capital enough to expend in bringing it into condition. One reason therefore why land covered with broad-leaved trees is universally valued is that, besides being for the most part really good at least on the surface, it will give a succession of abundant first crops by merely felling and burning the trees upon it, and scratching in the seed."

A similar mode of expression to that thus described of the American is in common use among the miners of Victoria. A "good poor man's diggings" is a phrase as familiar in the latter country as good poor man's land is in Canada or New Brunswick. By this expression the Victorian miner denotes an alluvial surface-deposit of gold, where the precious metal

---

* *Notes in North America*, vol. ii, p. 116.

can at once be obtained by the aid of the simplest tools and without any large expenditure of time or of money. These surfacings and shallow sinkings, as they are called, formed the earliest stage of mining industry in Victoria; and still constitute the great attraction in every " new rush." It was not for several years after the discovery of gold that abundant capital and efficient machinery began to produce upon the gold-fields their usual effects. Their introduction soon threw into the shade the feeble efforts of the unassisted miner. The alluvial diggings whose richness had at first attracted the attention of the civilized world, soon gave place to more fertile but also more costly mines. Their accessible stores were soon exhausted; and in point of absolute richness bear no comparison with the treasures more deeply buried in the earth, or with those embedded in the intractable quartz.

§ 8. When a man then has the power of choice between two natural agents, whether land or any other kind, of which one is intrinsically more potent than the other, if the means at his disposal enable him to use with advantage the superior agent, he will use it accordingly; but not otherwise. If his means be inadequate for its use, the inferior agent becomes for him at that time the superior. When the principle is thus stated, a reconciliation seems possible between the hostile schools of Ricardo and of Carey. Mr. Ricardo's theory, indeed, is rather partially true than absolutely false. It was constructed solely in reference to countries long previously settled, and well supplied with all the aids to production; and its author had not that familiarity with a different state of facts which only residence in a new country still in course of settlement can give. Mr. Mill declares* that the Ricardian theory does not and never was intended to apply to a

---

* *Political Economy*, vol. I. p. 519.

new country; an admission however which quite destroys its value as a general law. But on the other hand the two propositions which according to Mr. Mill are all that the Ricardian theory really involves, may well be accepted by those who concur generally in Mr. Carey's views. The measure of land-rent is the difference between the most fertile and the least fertile soil under cultivation ; and the cause of its payment is the necessity for simultaneously cultivating different qualities of land. In other words, rent is paid for the use of land, as it is for the use of any other instrument, when the cost of production upon two pieces of land varies; and the actual amount so paid is determined by the excess of the superior instrument above the average instrument. These and some other matters relating to the price of land will be more fully understood when the principles of exchange have been ascertained. Assuming however that rent is the result of a differentiation, it is plainly immaterial, so far as its payment is concerned, whether the difference be between the greater and the less, or between the less and the greater ; whether the order be from the more fertile to the less fertile, or from the less fertile to the superior soils. If the cost of production on two separate estates be different, this difference may arise not less by the diminution of the cost upon the one than by its increase upon the other. This difference Mr. Ricardo truly describes as rent. He has correctly explained both the nature of that source of revenue and the principles which determine its amount. But he unduly limits the circumstances which occasion the difference in cost ; and he excludes in favour of one cause of rent a second and more important cause. His affirmative propositions are true : it is his negations that his opponents are unable to accept.*

---

* See Mr. E. G. Wakefield's Note in his edition of the *Wealth of Nations*, vol. ii. p. 216.

§ 9. We have still to consider whether it be true that additional labour applied to the same land under the same conditions of agricultural skill gives a constantly decreasing return; and if so, whether the circumstance be peculiar to land. The proof of the existence of this tendency is the fact that occupiers of superior lands frequently bring under cultivation lands of inferior quality, instead of increasing their expenditure upon the better soil. It is found by experience that if the outlay upon any particular field be doubled, the doubling of the product does not follow. There is doubtless a point at which fresh applications of capital will cease to yield any return whatever. This statement amounts only to the assertion that the productive powers of any given portion of land are not unlimited. If they were, the whole condition of the world, moral as well as material, would be different from what it now is. There would be no motive to seek improvements in the art of agriculture or in its kindred arts and sciences. There would be no motive to occupy a greater surface of ground than that which would afford mere personal accommodation to the possessors. But as man's destiny is to replenish the earth and subdue it, he must be led to scatter himself over a wider extent of country; and he must have a sufficient stimulus to rouse into activity those mental powers which in the midst of the profusion of nature remain dormant or dwindle through inaction. The greater part of the earth therefore is slow to yield in any abundance her fruits; and man must either ascertain and comply with the terms on which only the full bounty of nature is granted, or must seek new scenes of industry. Thus the existence of this principle cannot be disputed; and we can to some extent trace its utility. But on the other hand, under this law man has sought out many inventions; and the condition under which alone the law is true, namely, that agricultural skill remains constant, is never in fact fulfilled. Although one of the earliest occupations which man has empirically exercised, agriculture is one of the latest of scientific arts. Its premises are derived from so many and such complex sciences that its advance must necessarily be slow. The theory of an art can never proceed more rapidly than the sciences upon which it is founded; and the sciences with which agriculture is more immediately connected are later in their development than those which give rise to purely material arts. For agriculture has to deal not only with mechanical and physical and chemical conditions, but with all the phenomena of life, and even to some extent with those of society. Montesquieu has well remarked that countries are not cultivated in proportion to their fertility, but to their liberty. There must be no inconsiderable degree of political knowledge where agriculture can be pursued in the most favourable circumstances. It is indeed strictly correct to say that the theory of agriculture depends even for its truth upon the state of the law. The mode of culture which in ordinary circumstances may be profitable, may become inexpedient when the freedom of commercial dealings is disturbed. The restrictive condition therefore upon production is continually kept in abeyance, although it still remains as it were *in terrorem*. Thus the aspect under which this law is of practical importance to us is not the steady tendency towards diminishing returns, and the counteracting influence, so far as it goes, of improvements. It would be more exact to say that improvements in agricultural skill are a condition precedent of any increased return. The control of man over nature which such improvements imply seems, so far as we can at present judge, to have no bounds. The time however may possibly come when all improvements have ceased, and the whole earth is fully inhabited; when there are no longer any waste lands fit for cultivation, or any further resources of art available to increase upon the existing soil the efficiency of labour. In such circumstances the progress of the human race in its present form will doubtless be arrested: but such a time and its consequences we may well leave to the overruling care of the Supreme Wisdom.

§ 10. But whatever may be its precise limitation, the law of diminishing returns cannot be held peculiar to land. There is no natural agent to which it does not in like manner apply. There is no limit either to the supply of land or to its productive powers in any other sense than that in which all other natural agents are limited. It is not the fact that the natural agent, land, is more limited in extent than any other natural agent. Land having peculiar degrees of fertility, or situated in particular localities that possess advantages as to the means of obtaining any of the usual aids to production, is limited; and so too are many other natural agents; but in no reasonable sense of the word can it be contended that the surface of the earth, without reference to any expenditure of capital upon it, or to any seats of population, is more strictly limited than the minerals beneath its surface, or the water upon it, or the air above it. No natural agent has or can have in any portion of it unlimited productive powers. The force contained in a given amount of electricity, or of gas, or of steam, is as rigorously limited as the vegetative powers contained in an acre of land. The comparison has generally been made between a particular portion of land, and some other agent to the quantity of which no limit is expressed; and not, as it obviously ought to be, between a specific portion of each. If we direct our attention to some such definite portion of any other natural agent, we shall at once observe that it presents the same phenomena as land. The quantity of sea-fish is generally supposed to be practically unlimited; yet experience has actually shown that the application of doubled capital will not give doubled returns. In the herring fishery for example, we are told that "the increase of the herrings taken bears no proportion to the extended netting."* In a given state of skill in the cotton, or the wool, or the silk manufacture, will the application of double the former capital to a pound of cotton, or of wool, or of silk, give twice the former product? Gold-leaf can be beaten out to an almost incredible fineness; but in the absence of any increase of skill or improved machinery, does the labour always continue proportionate to the result? If we double the labour expended upon a bushel of coals, can we thus generate double the quantity of steam? Or if we double the labour expended upon the steam that is produced from these coals, can we thus obtain double the former amount of force? If, to use a favourite illustration of Bacon, we double the food given to a hen, will she consequently lay double her previous quantity of eggs? Or is there any natural agent available for industrial purposes of which the product, while the industrial skill and its auxiliaries remain unchanged, can by the mere application of additional labour be made to increase in geometrical progression?

§ 11. The reason why in all these cases the tendency towards diminishing returns is not practically felt, is the same as that which explains the same phenomena in the case of land. The condition upon which the law of diminishing returns comes into operation is never realized. That condition assumes that the skill and the power of the labourer are unchanged. But the state of knowledge and of skill, and the

---

* *Report of British Association*, 1854, vol. xxiii. p. 134.

resources for aiding his labour at the disposal of the labourer, never do remain unchanged. In a healthy state of society, both agriculture and manufactures and every other branch of industry constantly tend to increase ; and the gross produce is consequently augmented. If the social organism be diseased, they are in the like manner diminished ; and the gross produce is consequently reduced. Since in the same state of skill and with the same amount of resources additional labour yields a return absolutely greater than before, but less in proportion to the cost, if the state of such skill or assistance be improved or be deteriorated, additional labour will yield in the one case a return both absolutely and relatively greater, and in the other case a return not only relatively but absolutely less than its former return. There is then no difference in this respect between land and other natural agents. If in certain circumstances there be a diminishing return from land in proportion to the labour expended upon it, there is in the same circumstances a similar diminution in the returns from other agents. If in different circumstances the earth yield her fruits more abundantly, other natural agents will under the like conditions give an increased product. When the facts as we daily see them are rightly understood, there is nothing in them to rebut the strong presumption in favour of this resemblance which the uniformity of nature in all other matters so evidently raises.

CHAPTER VII.

OF THE AIDS TO LABOUR.

§ 1. Since the efficiency of natural agents depends not only upon the character of the agents themselves, but also upon the circumstances of those who use them, the increase of their efficiency will not be limited solely by the extent of our discoveries and our improvements. The acquisition of an additional power to human labour may afford as great and as effectual a means of augmenting the utility to the labourer of natural agents, as the discovery of any new agent or any improvement in any existing process. The question therefore arises, what are the allies that in his warfare against external nature man has called to his aid? Weak as he is beyond all other of the larger inhabitants of this globe, he has subdued to his purposes, and will probably continue so to subdue, the fiercest animals, and the most violent elemental forces. If we enquire into the agencies by which so remarkable a result has been accomplished, we shall find that there are three ways, and three ways only, in which it is conceivable that man can reinforce his unassisted physical powers. He may since his constitution is complex help himself; or he may obtain help either from nature, or from other men. He may make his intelligence and his moral

powers supplement the weakness of his body; or he may
convert to his own use the active forces of nature; or he
may procure the aid of other men. From within, or from
without, and, if the latter, from either natural agents or other
men, such are the only earthly sources from which man can
invoke help.

§ 2. These sources of aid, however, are within man's
reach, and the assistance they have afforded him has proved
sufficient for all his purposes. He can by thought and
patience, by the vigorous use of all his faculties of mind and
of body, increase, as we have seen, the efficiency of his labour;
and so obtain an ever-increasing supply for his wants. At
first these conditions are very imperfectly observed; and his
wants are very ill supplied. He seldom exerts all his faculties; and he seldom rightly exerts those faculties which he
has developed. Yet still, whether from his own exertions,
feeble and ill-judged though they be, or from the bounty of
nature, he contrives to obtain something more than the actual
necessaries of existence. When he has obtained this surplus,
he has in his hands a new power. He has the means of
exerting his labour under more favourable conditions than
before. He has acquired sufficient means for his support
during a certain period which may be devoted to further
labour. It rests with him to determine whether he will
pursue his labour during that period; or whether he will
enjoy the repose which he has earned. His natural inclinations prompt him towards the latter course. But man is
not the mere creature of appetite. He can control his impulses, and forego a present enjoyment for a greater future
gain. He can foresee his advantage, and he can exercise
sufficient forbearance to attain that advantage. He is not
slow to perceive that for any extensive production something
further than the mere presence of the primary agents of pro-

duction is required. Industrial processes are often slow. Between the commencement of the work, and the enjoyment of its results, a lengthened interval frequently elapses. The instruments, too, which we require to assist us are themselves often the result of much time and labour. But our primary wants are urgent, and do not admit of any interruption to their supply. The labourer cannot wait for the completion of his work to obtain food and clothes and shelter. He must during its continuance obtain from time to time the means of support for himself and those who depend on him. He must possess the necessary materials for his work; he must be supplied with proper tools. All these essential requisites can only be provided from the accumulated results of previous industry. He therefore applies a portion of the fruits of his past labours to the accomplishment of labours of slower returns. That which he may have intended, and which he certainly had the power, to have used for his present enjoyments, he uses for his support or convenience while he is engaged in further production. The new product restores to him if he be ordinarily successful his former accumulation, and more besides; and thus at once enables and encourages him to persist in his temporary self-denial. Such is man's first and hardest struggle, the struggle with himself: a contest indeed that is never ending still beginning; but in which his first victory supplies him with his great engine for future triumphs, a scanty but still a veritable capital.

§ 3. In every attempt to satisfy human wants some effort intervenes between the desire and its satisfaction. This effort is in its very nature more or less painful, and ought consequently to be reduced to the least possible amount. It is therefore a clear benefit to mankind if any such effort or any part of it can be transferred from human beings to lower animals or to some inanimate force. The power to effect such a result arises from the nature both of the external world and of man. The objects which surround us afford materials not only to work upon but to work with. Some natural agents produce changes in other natural agents according to determinate and immutable laws. Man on the other side works both with his hands and with his brain. His labour is not only muscular but also nervous. He both is himself a motive power; and he is competent to ascertain the laws under which the forces of nature act; and to fulfil the conditions which their action requires. He can thus, if he take the proper means, make nature do his work; and can by mental effort substitute for the exertion of his own muscles the muscles of some lower animal or the force of some inanimate agent. It is this substitution of elemental powers directed by human intellect, for human muscles acting upon elemental capacities, that constitutes invention. The office of man in relation to the external world is, it has been truly said, that of the engineer. He wastes his power when he personally undertakes work which can be performed by inferior agencies; and these inferior agencies remain useless, or inert, until they receive their direction from the human intellect.

§ 4. But man is not dependent upon his solitary efforts, whether of body or of mind, or upon the assistance that by such efforts he can induce nature to afford. He is by his very constitution social: and he is accordingly able to assist, and to receive assistance from, his fellow-men. The actual constitution of the social unit, the family, and the difference of sex and of strength which it implies, and the protracted period of helplessness in childhood as compared with the similar period in the young of other animals, involve a certain amount of co-operation, and of co-operation shown as well by joint exertion as by separate occupation. The care of the children naturally devolves almost entirely on the mother. The supply of food, or the protection from danger, generally requires that activity and strength which none but a man possesses. The protracted duration of childhood induces habits of subordination and of sympathy which continue long after the age of childhood is past. When these feelings are cemented by a sense of mutual interest, the connection, so transitory in the lower animals, continues often in full force between the human parent and his adult children. Their labour is combined with his; and thus the family contains within itself the means of a constantly increasing co-operation. So natural indeed to man is the companionship with his fellow-men, that it is in the social state alone that his powers are developed. In the lower animals almost all their faculties seem to be independent of their situation. A reptile or a bird, a salmon or a dog, retain in their solitude the same powers and exercise them in the same degree that they possessed or exerted when they were in company with others of their kind. But the isolated man loses almost all his distinctive powers. Whatever he may previously have been, when he is abandoned, he quickly falls into a state of hopeless degeneracy. His physical powers are unequal to provide for the wants which he is capable of feeling. These wants associated industry is more than able to supply. There are many things which one man alone cannot do, but which are readily accomplished by the simultaneous action of several men. Two or three men when acting together can thus effect what a thousand men acting independently of each other would fail to perform. The raising of a given weight, for example, requires a certain force. This force is obtained when the power of two men is simultaneously applied. This force is never obtained as long as the labour of this or any other number of men is successive. There are other acts which, when performed by one person, are sufficient for the wants of several persons. The fire that is requisite for the uses of one man will equally suffice for the uses of several: and thus the labour and the material necessary for the kindling of separate fires will, if the men act together, be set free for other purposes. Two or more men, therefore, who are engaged in similar occupations will economize their labour, if by mutual agreement they severally undertake separate parts of the work, and share the results of their associated industry. By thus separating their occupations men are also enabled to obtain the full benefit of their respective powers. It is the diversity in the powers and opportunities that exists among different men which, at least when such diversity is conspicuous, forms the determining motive for their respective choice of occupations. When a man who is capable of more productive work undertakes labour of which the results are less productive, even though he execute each kind of labour equally well, he wastes his powers to the extent of the difference between the two classes. Thus when two men work together there is an obvious advantage, not only in each taking a different portion of work, but in each taking that particular portion for the performance of which he is best fitted. Such then are the immediate and obvious advantages of co-operation. It collects and unites the scattered elements of power; it economizes their use; and it organizes their action.

§ 5. Co-operation relates to the performance either of the same act, or of different parts of the same act. In the former case the combination of labour gives the greatest amount of power; in the latter case the division of employments gives the most judicious use of that power. But it frequently happens that two men without any previous communication have severally acquired some transferable objects; and that, when they are mutually aware of the facts, each prefers to his own his neighbor's property. In these circumstances the parties exchange their possessions. Each obtains, in return for something he comparatively disregards, something that he comparatively prizes; and each therefore gains by the transaction. The inducements to adopt this course continually and spontaneously suggest themselves. Men have an endless variety of tastes; their powers are hardly less varied; objects calculated to gratify these tastes, or within the range of these powers, are distributed in different quantities as well in adjacent as in widely separated localities. Men consequently are not slow to perceive that exchange is equivalent in its results to undesigned co-operation. But exchange, although it in some respects resembles the division of employments, is essentially a distinct operation. The two agencies differ from each other as a contract of partnership differs from a contract of sale. Exchange may be effected where the objects exchanged have been produced without previous concert; and it relates not to different parts of the same object, but to objects that are wholly distinct. Hence the influence of co-operation more directly affects the element of labour: while exchange not only economizes and organizes labour, but also renders the same service in respect to natural agents. As co-operation utilizes the diversity in human faculties; so exchange also utilizes the varying distribution of the forces of nature, and the different forms which in consequence of this diversity labour in different places assumes. Like co-operation, it economizes labour which might otherwise be wasted upon objects for the attainment of which the labourer does not possess sufficient natural advantages: like co-operation, it turns that labour to the greatest advantage by using to their fullest extent the means at the labourer's disposal; and by obtaining, in consideration of the surplus over his wants, the similar surplus possessions of his neighbor. Exchange therefore if it do not as between two persons add to their joint resources, so marshals and husbands these resources as to render them fully available, and to produce the greatest possible amount of advantage to each recipient. It is the result of an intelligent perception, quickened perhaps by desire, of the best means to a given end. The end is enjoyment. The means is effort. Exchange is in these circumstances the best means because it involves the least effort. It involves the least effort because we obtain, at the expense of something that we do not immediately or urgently want something which we could not otherwise procure, except, if at all, at a greatly increased cost. This advantage we obtain not at the loss but to the gain of the other party; because the distribution both of natural agents and of human powers is such that different men possess of different objects more than they severally desire; and because human wants are always in action and are always varied.

§ 6. The nature and the extent of the assistance that these agencies afford to industry will be considered in the following chapters. The fact that they do render such assistance is notorious. When two persons of equal abilities and in similar circumstances differ greatly in the amount of their respective possessions, to what cause is this difference due? The rich man is rich because he has accumulated a large capital; because that capital enables him to obtain the assistance of powerful natural agents; and because he can command the labour, or the products of the labour of other men. The poor man has none or few of these advantages. His capital is small. He is consequently unable to provide the costly machinery by which the powers of nature enlarge production. He cannot organize a staff of co-operators: and whatever may be his desires he has little to offer others in exchange. It is from his more complete control over these instruments, or some of them, that the large producer has generally an advantage over the small producer. The owner of a factory may not be even equal in the efficiency of his personal labour to a handloom weaver; but he has capital and the command of powerful natural forces to do his work; he can procure to any extent that he desires the co-operation of other men; and he has all the materials for conducting an extensive trade. Not only as between two contemporary fellow-citizens, but in different generations and in different nations, the action of these causes is discernible. If any one were asked why Englishmen of the present day are richer than the Hottentots now are, or than their own ancestors a thousand years ago were, his reply would doubtless be that it is because we have more capital and better implements than they; because our industrial organization is more complete; and because our commercial transactions are more extensive. The leading industrial characteristics then of an advanced community as compared with one less advanced, whether contemporaneous or preceding, are the greater accumulations of the former; the greater control which it exercises over natural forces; and its greater capacity for co-operation and exchange. It is in these respects so far as industry is concerned that we, like the son of the glorious Kapaneus, "boast that we are greatly better than our fathers;" it is in these respects that we hope and believe that the industrial triumphs of our sons will be superior to our own.

Nor is the influence of these agents confined to their direct action. They not merely assist labour but also affect its intrinsic efficiency. In theory, we may indeed conceive the efficiency of the labour apart from these auxiliaries. But in practice, it is only where all these auxiliaries are in operation that the conditions which determine the efficiency of labour are fulfilled. Education, if it be at all systematic, implies both capital and co-operation. Sanitary improvements cannot be effected without the assistance both of these agencies and also of inventions. No more potent stimulus to energetic labour exists than the opportunities of successful returns which exchange affords. Discoveries and improvements are not readily made without considerable outlay and much co-operation. As between themselves, too, these various agencies exhibit that remarkable reaction which so complicates all social inquiries. Capital depends upon the fund from which savings can be made; and this fund depends upon the extent to which the several aids have increased the efficiency of labour. Inventions can only operate under certain conditions; and these conditions require both capital and co-operation. In like manner both co-operation and exchange limit, and are themselves limited by, the other two aids and each other.

§ 7. It might have been expected that such advantages would have been quickly recognized; and that the objects which afforded them would always have been eagerly pursued. Yet the actual course of events has been very different. There is not one of these agents which has not been at some time generally neglected or disliked. With a strange perversity of judgment, men, as we sometimes can see only enemies in those who in reality are our surest friends, have attributed their sufferings to those very agencies which relieved them. The outcry against capital is still familiar to every ear; and the furious hostility with which socialistic writers regard it is far from being yet extinct. Even to this day an intense jealousy of this great agent prevails amongst the labouring classes of England; and multitudes are still persuaded that profits are a mere robbery, an unjust tribute exacted from the workers by those austere men who love to reap where they have not sown, and to gather where they have not strawed. These views, not indeed unsupported by unsound economic theories, have been sanctioned by still higher authority. Even in the present reign the acceptance of a tenth part of the profits by the owner of the capital when the capital itself was managed by another person was branded by law, as it once was by religion, with the odious name and the penal consequences of usury.

Still more intense than the hatred to capital has been the hostility to inventions. Upon the first introduction of the winnowing machine, the Scotch clergy denounced the attempt to produce an artificial wind as impious. So far did they carry their opposition that it is said that a person who had been guilty of raising "Devil's Wind" was on that ground refused admittance to the communion.[*] It was but a few years ago that an opposition, not indeed very strong or very lasting, was raised on the same ground of impiety to the use of chloroform, especially in obstetrical cases. When the first railway was projected the surveyors were mobbed in the field; and the engineers were ridiculed and browbeaten in the witness box. There has scarcely been one important discovery that has not met with virulent opposition. It would be useless to repeat the lamentations of philanthropists over the loss of employment to the poor which machinery occasioned; or to narrate the history of the violent riots by which the labouring classes sought to redress this fancied grievance. All these things belong to a time when the real influence of machinery was little understood. But even at the present day, and with all the experience of so many years, we find traces of the same feeling. One main cause of strikes has been the introduction of new

---

[*] Smiles's *Lives of the Engineers*, vol. ii. p. 106.

machinery. In 1859 a Liverpool ship builder got the copper for a ship's bottom punched by machinery ready for nailing on; but his workmen struck and obliged him to pay the hand punchers for going over the work as if it had not been already done. The shoemakers of Northampton struck against the use of the sewing-machine; and so strong was the sympathy of trade societies that subscriptions in aid of this strike were sent both from persons who themselves used in other localities these machines, and from the very house in London which supplied the obnoxious articles.* Yet there have been few more decisive instances than that of sewing-machines of the impulse which a useful invention gives to production, and the extent to which wages are raised by the increased demand.

In uncivilized or partially civilized countries, that is to say, in by far the greater part of the world, co-operation in any other than a mere rudimentary form is almost unknown. Even when circumstances are favourable, and it has begun to appear, both law and custom which supports law regard with jealous dislike every tendency to union. In France commercial associations were permitted to exist only by royal favour; and this favour was seldom extended unless the applications were supported by powerful interest or by other arguments still more potent. So lately as the present century the formation of joint associations with transferable stock for any purpose was by statute absolutely prohibited in England. Lord Eldon intimated his opinion that companies with large capital arising from numerous small contributions and with transferable shares were injurious to the public, and were consequently illegal by the common law.† So strong indeed in this

---

* *Journal of Statistical Society*, vol. xxiv. p. 500.
† See 1 *Lindley on Partnership*, p. 150.

direction was the tendency of the courts during the earlier part of this century that it seems strange how such companies escaped from a distinct judicial prohibition. It is only within the last few years that the law relating to associations has been placed on anything like a satisfactory footing.

The dislike thus felt towards co-operation is in early societies still more strongly shown towards exchange. The homogeneous structure of early society, in which separate families formed a number of similar and self-sufficing associations, and a kind of plurality of ownership in the family property among the members of each family prevailed, was highly unfavourable to the combined action of several families, or to the interchange of their several possessions. In ancient Persia accordingly exchange, when it was at all permitted, was confined to the humbler classes exclusively. In a much more advanced state, the Romans held all trade to be dishonourable. Similar sentiments prevailed amongst the mediæval aristocracies. Even in our days of modern enlightenment the utility of exchange is far from being at least in practice recognized. It is not two hundred years since the English Parliament declared the trade with Ireland and the trade with Scotland to be nuisances. It was at the very close of the last century that the Chief Justice of England judicially declared that an agreement to buy from certain hopgrowers all their hops during the following year at a fixed price was "an evil of the greatest magnitude, a most heinous offence against religion and morality, and against the established laws of the country."* Our statute law has indeed imposed innumerable restrictions upon exchange whether between citizens or between citizens and foreigners. Slowly and with difficulty have

---

* *Rex v. Waddington*, 1 East, 166, per Lord Kenyon, C. J.

these restrictions been abolished: still more slowly and painfully are some of the leading European states following our example; while in the new world free men are deliberately fixing upon their industry the fetters that Europe has so struggled to burst. So blind are men to their real interests, and so true is the Roman satirist's remark that Heaven by granting our prayers would most effectually seal our doom. Ignorance, that arch-enemy to our well-being, is continually striving to pluck by a sad perversion the nettle Danger from the flower Safety.

CHAPTER VIII.

OF CAPITAL.

§ 1. Capital according to the description usually given of it consists of the materials upon which the labourer operates, the tools and appliances of his work, and his needments during the period of production. None of these things are labour, or the results of labour alone. None of them are natural agents, unaffected by any human interference. They are the products of labour and of natural agents. They are the results of previous industry which the owner applies, not to immediate enjoyment, but as means to procure additional sources of enjoyment. Capital then consists of commodities that have been saved and are used for further production. It consists of commodities, distinguished from services in the more restricted sense of that term; since it is the result not of labour alone, but of labour exercised upon and embodied in natural agents. The commodities are saved, for the owner, while it was in his power to use them in any way he might desire, has by an exercise of self-denial abstained from his present gratification; and has postponed his immediate to an increased future enjoyment. Thus capital springs from a moral act done towards a material product. The motive from which this act proceeds supplies another essential part of the definition of capital. It is not every saving of commodities that constitutes capital. The commodities thus saved must be used. But this use always requires qualification. Commodities that have been saved may be used for mere enjoyment without any ulterior purpose. In such circumstances they form a part of what economists usually describe as revenue. That which characterizes capital is its application to the purposes of production. It is the will of its proprietor, not any intrinsic quality, that converts a mere hoard, a quantity of saved commodities, a possible portion of revenue, into capital. Hence the same commodity may be and often is at different times capital, and not capital. The bullock which when living formed part of the capital of the grazier, and when dead, of the butcher, is not capital when the meat reaches the consumer. Capital is invariably a means to a given end. It is an instrument which facilitates the accomplishment of a certain result. The great end of all human industry is enjoyment, of whatever kind, and at whatever period. One principal means by which an increase of the instruments of this enjoyment can be attained, one great agent which aids industry in its work, is that effort of the will which, sacrificing the present to the future, converts wealth into the fruitful parent of multiplied wealth.

§ 2. "In the order of history," says Bentham,* "labour precedes capital: from land and labour everything proceeds. But in the actual order of things, there is always some capital already produced which is united with land and labour in the production of new values." This capital must in every case have been itself the result of some previous capital. Agriculture and the spontaneous bounty of the earth alike

---

* *Works*, vol. iii. p. 36.

furnish their supplies only at stated seasons of the year. The labour of the hunter and of the fisher is perhaps the only labour which yields an immediate return; and hunting and fishing cannot be successfully pursued without some kind of tools, however humble. Materials even in the most ordinary kind of manufacture are generally themselves the finished products of anterior workmen. When labour is directly exerted upon natural agents, as upon stone or timber, it is seldom possible to use the object on the spot on which it was found. For the preparation and the removal of these things some aid is required. Man cannot dress stone and fell timber with his bare hands, or carry loads long distances upon his back. He must have capital. But other forms of capital were necessary for the formation of this capital; and these again imply the existence of others more remote; and so we ascend from difficulty to difficulty until our speculations are lost in darkness. How then did capital begin? If its presence be essential to industry, how could industry have originated? To such questions there are no means of reply. The earliest records of our race imply the existence of capital. The expressive testimonies which modern linguistic science has discovered in the primitive words by which our Arian forefathers long before the age of written record expressed their thoughts and feelings, all recognize the presence of accumulation.* There is no savage tribe in which we cannot trace its rudiments; and yet even with such aids as they are known to possess, we have no instance of the spontaneous development of the savage into the civilized man. If any such case has occurred, we cannot now recover any authentic vestiges of the process. All that is known is that as far as the evidence extends, men have never been without some accumulation: that whenever that accumulation becomes

---

* See Professor Max Müller on the *Science of Language*, p. 223.

considerable, its rate is accelerated; and man's general well-being is increased: that on the contrary, when the amount is decreased, the difficulty of accumulation is augmented; and man steadily tends to degenerate. It must have been in its infancy therefore that industry most required aid; and it must have been in its infancy that the necessary aid was most difficult to procure. In the moral as in the physical world, we can trace the operation of laws which sufficiently explain the continuance of the present system; but the origin of that system is beyond our unassisted powers.

§ 3. It is difficult to estimate with any precision the amount of the assistance which capital renders. This difficulty arises partly because the operation of capital in a great degree consists in facilitating the use of the other industrial auxiliaries; and partly because its direct effects are habitual rather than immediate. Many a man has commenced life at a daily wage of a few shillings, and has afterwards counted his income by tens of thousands of pounds. The efficiency of such a man's personal labour may not have been much less in the time of his poverty than in the time of his wealth. It was the amount of assistance which his labour received that was different. But a large part of that assistance consisted of invention, co-operation, and exchange, which although they may have been obtained by capital are nevertheless distinct from it. In like manner the absence of capital not merely involves the want of its own peculiar assistance, but prevents the use of the other aids; and so retards the growth of wealth more than the loss of its own advantages would of itself occasion. It would be possible perhaps to ascertain the quantity of labour saved in any case by the continuity which capital gives to it; but no one could estimate its influence upon the skill or the habitual energy of the worker. But although these influences do not conveniently admit of numerical expression, they are not on that account the less certain or important. The half-caste Indians often show great energy and skill, and have European arms and exhibit considerable powers of co-operation; but from their improvidence they are often after the hunting season is over reduced to the extremity of want. Even the skilled operative of Lancashire, although he is acquainted with inventions, and is accustomed to co-operation, and although an eager market for his products awaits him, is, if the use of the capital that employed him be withdrawn, left hopelessly idle.

So powerful is this influence of capital that economists have generally described it as one of the primary agents in production, or at least as holding equal rank with those agents. It is indeed difficult to imagine how, without some accumulation, man, such is his physical weakness, could provide even the scantiest means of subsistence. As a question of fact, the lowest savages do not seem to be altogether destitute of what may be called the accumulative faculty. But this constant presence of capital ought not to obscure its real position or its actual part in our industry. Each of the other industrial aids appears to be not less indispensable than capital. If it be true that at no time of which we have any record, and among no people with whom we are now acquainted, do we hear of men living without some capital, it is no less true that we have no knowledge of men living without some attempt to avail themselves of the motive forces of nature, or without in some degree combining their efforts, or without exchanging with their companions the results of their labour. At no time and in no place are any of these aids wholly wanting. The state in which they are found is sometimes indeed merely rudimentary; but still their existence can always be clearly discerned. So strongly has this been felt that each of these aids have, like some other phenomena of our nature, been in turn described as forming the essential characteristics of man. They are all undoubtedly peculiar to man, although they are not all distinctive. They result from certain intellectual and moral faculties, according as these faculties, in a greater or less degree, but always in some degree, are called into action by surrounding circumstances. None of them can therefore be regarded as a primary force; and as between themselves the only difference seems to be that capital usually precedes the rest. The operation of the four is alike. They are neither labour nor natural agents; but contrivances for increasing the utility of both. By their mutual reaction, as we shall subsequently see, they greatly increase each other's power; and none of the four can attain its full efficacy while the development of any one of the rest is incomplete. It seems therefore the proper course to rank all these aids together as auxiliaries; and it is needless to embarrass a subject already sufficiently intricate by raising unprofitable questions as to their relative importance.

§ 4. The first and most obvious mode in which capital directly operates as an auxiliary of industry is to render possible the performance of work which requires for its completion some considerable time. In the simplest agricultural operations there is the seed time and the harvest. A vineyard is unproductive for at least three years before it is thoroughly fit for use. In gold mining there is often a long delay, sometimes even of five or six years before the gold is reached. Such mines could not be worked by poor men unless the storekeepers gave the miners credit, or, in other words, supplied capital for the adventure. But in addition to this great result, capital also implies other consequences which are hardly less momentous. One of these is the steadiness and the continuity that labour thus acquires. A man, when aided by capital, can afford to remain at his work until it is finished, and is not compelled to leave it incomplete while he searches for the necessary means of subsistance. If there were no accumulated fund upon which the labourer could rely, no man could remain for a single day exclusively engaged in any other occupation than those which relate to the supply of his primary wants. Besides these wants, he should also from time to time search for the materials on which he was to work. The quantity of his labour would consequently be diminished; and desultory and irregular habits would be formed. Thus, both the available period for work would be reduced, and there would be wanting that habitual energy and steady persevering industry on which the efficiency of labour mainly depends. Again, it is to capital that we owe that condition of things in which the improvement of labour is possible. Without its aid there would be no opportunity of general education; no opportunity of combining and generalizing those results of experience which give rise to art; no means of acquiring that knowledge of external nature which is the only safe basis of action: no means of carrying into effect those sanitary arrangements which are essential to health and longevity. Capital thus places labour in a condition favourable to its efficiency, and secures the means of improving it.

§ 5. Capital, as we have seen, is the consequence of a triple antecedence. It is the result of forbearance, of labour, and of natural agents. When we have produced certain commodities, we forbear to apply to our immediate gratification the entire product. It is the surplus of income over expenditure, whether such surplus arises from increased production or from diminished outlay, that is meant by saving. This surplus thus saved, when applied to the production of other objects, increases capital. But although capital by an act of forbearance is actually saved, it is saved only to be spent. The commodities that have been saved are consumed in reproduction. Capital is not, as our modes of expression lead us to think, wrapped up in a napkin, and buried in the ground. It is employed in active business, that it may be replaced with an increase. The real difference between capital and revenue is not in the actual withdrawal of commodities from consumption, but in the motive for and the direction of consumption. In the one case, the consumption is the end; in the other, it is the means. But in both cases alike, whether nothing further be intended or desired, or whether a new and increased product be sought, the commodity is consumed. In the former case, it has discharged its duty, and is heard of no more. In the latter case, another commodity, not necessarily of the same kind, but of greater worth, generally supplies its place. "Capital," as Mr. Mill observes,* " is kept in existence from age to age not by preservation, but by perpetual reproduction : every part of it is used and destroyed, generally very soon after it is produced, but those who consume it are employed meanwhile in producing more. The growth of capital is similar to the growth of population. Every individual who is born dies, but in each year the number born exceeds the number who die : the population therefore always increases, although not one person of those composing it was alive until a very recent date."

§ 6. Some writers, from their deep sense of the importance of capital in promoting production, have eulogized that mode of expenditure which aims at some ulterior return, and have deprecated that other mode which rests in the

---

* *Political Economy,* vol. i. p. 92.

mere enjoyment of the thing consumed. The illustrious author of the *Wealth of Nations* himself, in this as in some other instances, misled by the subtle influence of eulogistic and dyslogistic terms, through his anxiety to avoid a popular fallacy of his day, fell into the opposite error, and pronounced the prodigal to be a public enemy and the frugal man to be a public benefactor. It has been the fashion since his time, partly from the weight of his authority and still more from the great demand for capital consequent upon the vast field of employment that has been in the present century opened to it, to magnify capital as the one thing needful either for individual or national prosperity. We thus lose sight of the fact that capital is but a means to another means. It is the servant of a servant. It is not in itself our object; but it is a means to enable our labour to procure that object. The sole object of labour is enjoyment: the sole use of capital is the assistance that it affords to labour in procuring that enjoyment. We save much that we may expend more. A man is the richer and as far as wealth is concerned the happier not for what he can save but for what he can spend. But while he saves that he may spend, he must also remember that if he wishes to continue to spend he must save. To praise capital therefore at the expense of revenue is to set up the means above the end. If we were to retrench all our unproductive expenditure, as it is somewhat unfairly termed, if we were to accumulate to the full extent of our power, it would seem reasonable to suppose that we should have at our disposal for all industrial purposes a far larger sum than we at present have. Yet this increase would not be permanent. When the gratifications which capital supplied were taken away the motive for saving would cease to exist: capital is but an auxiliary of labour; and labour is never expended save in the hope of procuring some gratification. Unless therefore we assume that men will continue to produce and to accumulate commodities without any adequate motive, the excessive capital so accumulated would not be replaced. If our wants were by any means reduced to the lowest point consistent with the preservation of life, our means of satisfying these wants, or in other words our wealth, would speedily undergo a corresponding decrease. It is not the repression of our wants, but their diversity and their extension, that is the condition most favourable to the abundance of wealth, and to the accumulation of those instruments that are suitable for its production. Such a restriction of enjoyment to the gratification of the mere primary appetites could never take place in man. It is characteristic not of him but of the lower animals. Sometimes indeed we do see a human instance of this kind, but it is only in the morbid acquisitiveness of the miser. No man in the ordinary state of mental health, far less any considerable number of men, will for the mere sake of saving deny themselves, when they have the power of enjoyment, all that makes life pleasant.

§ 7. It also follows from the proposition that capital is an aid to labour, that as men never save all they can, so they never spend all they possess. When accumulation has commenced, men soon feel that the more, within certain limits, that they save, the more they will be able to expend. The greater returns of assisted as compared with unassisted labour admit of at once a larger saving and an ampler outlay. Cases indeed occur where the wealth that might have been saved is wasted in folly or in profligacy. Sometimes also we find the acquisitive faculties abnormally weak. All these cases however are exceptional. Where there is a number of persons, the shortcomings of some are compensated by the excesses of others; and it is only to the average man that general principles will apply. Experience abundantly shows what even without its aid we might have expected, that when the conditions are favourable to its accumulation, there will never be a want of capital. "The principle," says Adam Smith,* "which prompts to expense is the passion for present enjoyment, which though sometimes violent and very difficult to be restrained, is in general only momentary and occasional. But the principle which prompts to save is the desire of bettering our condition, a desire which, though generally calm and dispassionate, comes with us from the womb, and never leaves us until we go into the grave. In the whole interval which separates these two moments, there is scarce perhaps a single instance in which any man is so perfectly and completely satisfied with his situation as to be without any wish of alteration or improvement of any kind. An augmentation of fortune is the means by which the greater part of men propose and wish to better their condition. It is the means the most vulgar and the most obvious; and the most likely way of augmenting their fortune is to save and accumulate some part of what they acquire either regularly and annually or upon some extraordinary occasions. Though the principle of expense therefore prevails in almost all men upon some occasions, and in some men upon almost all occasions, yet in the greater part of men, taking the whole course of their life at an average, the principle of frugality seems not only to predominate but to predominate very greatly."

§ 8. The moral influence that capital, both in its formation and its action, exercises upon the person that accumulates it, is very noteworthy. Experience soon teaches man the necessity of reinforcing his first feeble efforts to satisfy the imperious cravings of his nature. The acquisition of capital, the first and most obvious means for his purpose, not

---

* *Wealth of Nations*, b. ii. ch. 3.

only exercises the intellect by suggesting the adaptation of means to ends; but teaches the far harder lesson of forbearance. Man quickly finds that if he wishes to enjoy himself to any considerable extent, he must prepare for his enjoyment by self-denial. He must learn to put a check upon his present impulses, and to brace his mind to a steady contemplation of the future. The very exigencies of his position ensure, when the process of saving has once fairly begun, that predominance of the past, the distant, and the future over the present which, according to a well-known maxim, elevates a man in the rank of thinking beings. Again, the power to save is exerted not for the sake of the person himself only, but also for the sake of others. Apart from those other instances to which reference has been already made, the records of the Saving Banks show that savings are often effected by the very poorest people with the avowed object of assisting their parents, or some other near relative. The possession of the accumulation gives the power of doing kindly acts; and the sympathies thus excited stimulate to the acquisition of further means of beneficence. Thus the disposition to save is at once strengthened by, and itself strengthens, the charities of domestic life. It must also be remembered that the capitalist is by the nature of his property bound to promote order and peace. To his interests all disorder and all wars are peculiarly destructive. The larger the amount that a man has accumulated, and the larger the portion of that stock that he is induced to invest as capital, and the more distant the returns that he expects to derive from his investment, so much the more numerous and more binding are the pledges that he thereby gives to fortune: and so much the closer is the tie which binds him up with the settled order and the continuous improvement of the society in which he lives. Nor should we forget that the possession of capital implies also the possession of leisure:

L

and is a condition precedent to the cultivation of all those arts that refine and adorn life.

"Capital," says M. Bastiat,* " has its root in three attributes of man, foresight, intelligence, and frugality. To set about the creation of capital we must look forward to the future, and sacrifice the present to it; we must exercise a noble empire over ourselves and over our appetites; we must resist the seduction of present enjoyments, the impulses of vanity, and the caprices of fashion, and of public opinion always so indulgent to the thoughtless and the prodigal. We must study cause and effect, in order to discover by what processes and by what instruments nature can be made to co-operate in the work of production. We must be animated by love for our families; and not grudge present sacrifices for the sake of those who are dear to us, and who will reap the fruits after we ourselves have disappeared from the scene. To create capital is to prepare food, clothing, shelter, leisure, instruction, independence, for future generations. Nothing of all this can be effected without bringing into play motives which are eminently social; and what is more, converting these virtues into habits. Thus in whatever point of view we place ourselves, whether we regard capital in connection with our wants which it ennobles; with our efforts which it facilitates; with our enjoyments which it purifies; with nature which it enlists in our service; with morality which it converts into habit; with sociability which it developes; with equality which it promotes; with freedom in which it lives; with equity which it realizes by methods the most ingenious; everywhere, always, provided that it is created and acts in the regular order of things and is not diverted from its natural uses, we recognize in capital what forms the indubitable note and stamp of all great providential laws, harmony."

---

* *Harmonies*, p. 186.

CHAPTER IX.

OF THE CIRCUMSTANCES WHICH DETERMINE THE EXTENT OF CAPITAL.

§ 1. We have seen that capital consists of commodities that have been saved, and are applied to aid some further exercise of our labour. Its amount must therefore depend first upon the quantity of commodities that are at the disposal of the capitalist; next upon the strength of his inclination to save; and lastly, upon the strength of his inclination to employ his savings. The formation of capital, in other words, depends upon the ability and the will of its owner. His ability is regulated by the means that he possesses from which he can save. His will is determined by the strength in him of the motives which induce him to save, and by the strength in him of the motives which induce him to expend his accumulations in aiding his labour. For it must be remembered that it is not all wealth, or the power of acquiring wealth, that constitutes capital. The one may become capital, and the other may be instrumental in its production; but in themselves they are different from it. The name capital must be reserved for that portion of wealth which is appropriated, whether by way of maintenance or of mechanical aid, to the assistance of labour.

§ 2. The first condition on which the amount of capital depends is the ability of its owner to save. This ability arises from and is limited by the productive powers of his industry. The ampler a man's possessions, the greater is the amount that he can retain from daily use. The poor man can spend little, and can save little; the rich man can spend much and at the same time save much. The struggling clerk who with difficulty maintains a respectable appearance, finds it hard to pay the premium upon his policy of insurance. His wealthy employer who denies himself no reasonable indulgence, finds at the end of the year no inconsiderable portion of his income still unexpended. In no country and at no time has there been a greater expenditure of revenue than in England of the present day. In no country and at no time has the annual savings of any population reached an equal sum. But the productive powers of industry depend upon the efficiency of labour, upon the excellence of the objects upon which that labour is exercised, and upon the assistance that it receives from its several aids. Everything therefore that promotes the efficiency of labour or of natural agents, or of their auxiliary agents, tends to increase the productive powers of industry; and so to increase both the sources of capital and also as will presently more fully appear the objects of exchange. Thus the action of these two aids is both reflex and reciprocal; and each of them is in its turn affected by the operation of the other aids. The greater the amount of the aid, the greater if other things be equal is the assistance that it gives; the greater its assistance, the larger will be the product; the larger the product, the ampler will be the funds from which new capital may be accumulated.

§ 3. But assuming that the fund whence savings can be made exists, there are still great differences in the degree of saving that is effected by different persons, or even by the same person at different times. The motives to save are unchanged, but they frequently operate in different circumstances with very different degrees of force. The advantage may in fact be attainable, but the party interested may not be willing to pay the necessary price. We have therefore next to inquire into the circumstances which induce men to pay this price, or, in other words, into the conditions of the effective desire of accumulation.

Since the motive for saving is the prospect of some future advantage, several conditions must be fulfilled before the motive begins to operate. The advantage must exist; its existence must be fully perceived; and its sufficiency and the probability of its attainment must be acknowledged. The person who is about to make the sacrifice must be sure first, that an advantage may be really obtained; and next, that when it is obtained it will be sufficient to recompense him for the sacrifice he had made to secure it. Apart therefore from the power to save and the prospect of gain, the amount of accumulation will depend upon the intelligence and the foresight of the labourer, and upon the due subordination of his self-regarding desires, and the preponderance of those desires which Adam Smith comprises under the general designation of a desire to better his condition.

§ 4. The inducement to the self-denial which the accumulation of capital implies, is the hope of some future advantage. There are however many minds which are quite incapable of realizing any event that is even slightly remote. Their perceptions are so sluggish that they require the stimulus of the immediate presence of an object of desire. Their mental vision is so defective that an object which does not immediately come within the reach of their senses, or of which they have not had repeated experience, makes no impression, or at least no distinct impression upon them. As in the case of physical nearness of sight, such persons may and often do thoroughly realise an object which is placed sufficiently near them. In these circumstances their efforts to obtain it are as energetic as those of other men. But as the object recedes, the impression on their minds grows less vivid; and their sense of its desirability less intense; until at length it becomes indistinct, and fades altogether from their view. In such circumstances they cease to desire it because to them it practically has no existence. Hence arises that extraordinary improvidence which is the most striking characteristic of barbarous tribes; and which among civilized men is found in a greater or less degree in children, in slaves, and in those classes of free men who are least removed from barbarism. This improvidence is the result of the absence of those habits of perception and action which produce association of the present and the future, and of the intervening circumstances that connect these two periods. In the case of children, their habits have not yet been formed, and their acquisition of such habits constitutes one of the most important results of education. In the case of slaves and savages the want of these habits is a mental defect arising from the circumstances in which they are placed. The slave who at night does not know what he may be required to do in the morning, and whose entire earnings are the property of another, is little likely to attempt to regulate a future over which he can exercise no control, or to deny himself an enjoyment for the benefit of his master. The condition of savages is hardly less precarious, although in a different way; and their insecurity is hardly less complete. Savages are generally hunters; and their hunting grounds are occupied in common by their tribe. The acquisition of a better hunting ground is the act of the tribe not of the individual; and the possessions of the individual are limited by the quality of the hunting ground, that is, by the abundance of the animals that the ground so occupied by his tribe contains. The future therefore presents to him little that he can foresee or control. Such a state of things gives rise to superstition; and the superstition of the savage reacts upon his conduct, and prevents him from attempting to regulate even those contingencies that are within his powers. The North American Indians remarkably illustrate these principles. These people occupy extensive reserves of fertile land which the American Government has set apart for their use: they are not averse to labour where their reward is immediate; they possess some knowledge and some tools; and they are capable of doing thoroughly the work that they undertake. Yet they are constantly in extreme distress; they never exercise the commonest handicrafts which they see the white men practice; and they appear unable to appreciate the motives that lead the latter, while in possession of abundance, to incur increasing toil for a remote and uncertain futurity. " They thus afford," says Mr. Rae,* "a striking instance of the effects resulting from a great deficiency of strength in the accumulative principles. They have skill adequate to the formation of instruments capable of ministering to the necessities and comforts of a numerous population, for with the powers of fire, the axe, and the hoe, the great agents in converting the forest into the field, they are well acquainted; they have industry, content with a very moderate if immediate reward; yet from inadequate strength of this principle these all lie inert and useless in the midst of the greatest abundance of materials: and the means of existence in the time to come not being provided, as what was future

---

* *New Principles of Political Economy*, p. 138.

becomes present, want and misery arrive with it, and these tribes are disappearing before them."

§ 5. But the condition of accumulation which we are now considering is not confined merely to our intellectual development. It also includes the state of our social and sympathetic feelings. Men will often submit to privations for the sake of those dear to them, that no hope of personal gain would induce them to undergo. So far as the individual is himself concerned, a future good as compared with a present good is uncertain in its arrival, and may be inferior in its enjoyment. If therefore provision were made for a continuance of his wealth during his life, no man who regarded his own advantage alone and not the interests of any other person would deny himself any indulgence for the sake of saving. This intensity of the force with which the sympathies of our nature lead men to save, may be estimated from the contrast of this result of calculating selfishness with the facts that surround us. We see the sacrifices that men daily make to provide for or to assist their children. We see the great sums paid as premiums on insurances on the life of the person paying them. We know the sums contributed to our splended public charities; and we do not know the amount of that charity that is given in secret. On the other hand instances are not wanting of mere self-regarding providence, and of indifference to the interests of survivors. Annuities are often granted terminable upon the death of the purchaser. Such an expenditure of capital is equivalent to its conversion into revenue; but is intended to guard against the danger of the owner outliving his money. In the ordinary case of life-assurance a man saves from his revenue for the sake of others. In the purchase of annuities the man who has no interests or none that he desires to promote beyond himself, spends as revenue what might have been used to assist labour. The connection between selfishness and profusion was illustrated beyond all other periods in the time of the Roman Empire. In those evil days men did not desire to have children; and saw no blessing in leaving the rest of their substance to their babes. To be free from every domestic tie and to consume in their personal gratification all their property, were in their estimation much more reasonable aspirations. The man that is thrifty for the sake of his heir, Horace tells us, scarce differs from a madman. The poet writes as if his maxim required no proof and admitted of no dispute; and in the times of the later Empire its truth would appear still more self-evident. The subjects of the Cæsars cared neither for private ties, nor for public advantage. Their only aim was to acquire wealth, by fair means if it might be, but if not, by any means; and when acquired to squander it in every form of sensual indulgence.

There is however nearer home abundant evidence of the tendency to accumulate when the moral and social feelings are strong, and of the tendency to profusion when those feelings are weak. Almost all that large sum which constitutes the life-assurance fund of the kingdom is contributed by persons belonging to what are called the middle classes. But it has been shown that the families of factory operatives in Lancashire were recently earning higher incomes than many of the professional classes of England, higher, it is said, " than the average of country surgeons, higher than the average of the clergy of all denominations, much higher than the average of the teachers of the rising generation, and perhaps higher average of middle classes of the United Kingdom generally."[*] Similar high rates of wages prevail in those districts which are the seats of the

---

[*] Smiles's *Workmen's Earnings, Strikes, and Savings*, p. 22.

woollen manufactures. In the iron trade it appears that the rail rollers earn a rate of daily pay equal to that of lieutenant-colonels in Her Majesty's Foot-guards; shinglers equal to that of majors of foot; and furnace men equal to that of lieutenants and adjutants.* Yet the accounts given by credible witnesses of the improvidence of these classes read rather like the descriptions of savages than of prosperous and industrious labourers in the very centre of civilization. Higher wages only seem to bring to them additional means for low gratification. It is not they who use Savings Banks, or who support Mechanics' Institutes, or Schools. All that they earn in one week is spent in the next; and the whole population of great districts oscillates from year's to year's end, between alternate plethora and destitution. But it is not so much the recklessness of their expenditure as its objects that affects the present generation. On the latter subject, the late Mr. Porter, a very competent authority, makes the following remarks, "It has been computed that among those whose earnings are from 10s. to 15s. weekly, at least one half is spent by the man upon objects in which the other members of the family have no share. Among artizans earning from 20s. to 30s. weekly, it is said that at least one third of the amount is in many cases thus selfishly devoted. That this state of things need not be, and that if the people generally were better instructed as regards their social duties it would not be, may safely be inferred from the fact that it is rarely found to exist in the numerous cases where earnings not greater than those of the artizan class are all that are gained by the head of the family when employed upon matters where education is necessary. Take even the case of a clerk with a salary of £80 a-year, a small fraction beyond 30s a-week, and it would be considered

---

* Ib., p. 26.

quite exceptional if it were found that anything approaching to a fourth part of his earnings were spent upon objects in which the wife and children should have no share." So too, Mr. Clay analyzed the expenditure of 131 workmen employed by the same master, from statements furnished by themselves; and found that twenty-two per cent. of their aggregate wages was expended upon drink. This sum represented premiums upon insurance policies to the amount of £500 upon the lives of each of these men.

A very remarkable contrast to this carelessness is found in the habits of the Irish peasantry. Of this people it has been truly observed that whenever they are able to get good wages they never forget their relatives and friends who are in want. Much has been said and written of the ludicrous improvidence of the Irish: and there can be no doubt that in the absence of what he considers an adequate motive, the Irish peasant, like every other human being, will not sacrifice the present to the future. But the peculiarity in his national character is, that to the Irishman the great inducement for saving is not to increase his own personal gratification, but to benefit his friends. This phenomenon, whatever may be its ultimate cause, has been observed in every variety of place and circumstance. "The Irish street folk," says Mr. Mayhew,[*] "are, generally speaking, a far more provident body of people than the English street-sellers. To save, the Irish will often sacrifice what many Englishmen consider a necessary, and undergo many a hardship. Some of the objects for which these struggling men hoard money are of the most praiseworthy character. They will treasure up halfpenny after halfpenny, and will continue to do so for years, in order to send money to enable their wives and children and even their brothers

---

[*] *London Labour and Poor,* p. 115.

and sisters when in the depth of distress in Ireland to take shipping for England. They will save to be able to remit money for the relief of their aged parents in Ireland. They will save to defray the expense of their marriage, an expense the English costermonger so frequently dispenses with, but they will not save to preserve either themselves or their children from the degradation of a workhouse. On all the kinds of loan with which the poor Irish are aided by their countrymen no interest is ever charged."

For many years the Irish settlers in America had been in the habit of remitting large and constantly increasing sums to their friends in the old country. But although the aggregate amount was considerable, yet as each remittance was small, and sent through private channels, the fact did not attract public attention. At length on the occasion of the Irish famine, Mr. Jacob Hervey, a benevolent member of the Society of Friends, who like so many others of the same society took an active share in the attempts made to mitigate the effects of that calamity, made some inquiries as to the amount of their remittances. He estimated that in the year 1846 the remittances through the business houses of New York, Philadelphia, and Baltimore, amounted to about £200,000. "This," he says, "is sufficient to astonish everybody who is not aware of the fact, and it is but right that credit should be given to the poor abused Irish for having done their duty. Recollect that the donors are working men and women, and depend upon their daily labour for their daily food; that they have no settled income to rely upon, but with that charming reliance upon Divine Providence which characterizes the Irish peasant, they freely send their first earnings home to father, mother, sister, or brother. Contrast the sums subscribed by the rich to assist the poor Irish and they sink into insignificance." "When I recollect," he adds, "that these remittances have been constant for over five and twenty years past, increasing each year, I am at a loss what amount to put down as remitted in that time, but I should certainly say at least three millions of dollars."* Subsequently more extended inquiries were made by the Emigration Commissioners. From their report it appears that from the year 1848 to 1855 these remittances exceeded eight millions three hundred thousand pounds, or upwards of a million as the average of each year.† From Australia also, similar remittances are constantly made, although there are no means of ascertaining their exact amount. But in the year 1861, a sum of money was voted by the Parliament of Victoria for assisting immigration by paying a portion of the passage money for persons for whom the remaining portion was paid by some colonist. Within a week applications for considerably more than the whole sum were made by the Irish alone : and to prevent such an absorption, it was thought expedient to divide the sum according to the separate nationalities.

§ 6. The amount of accumulation thus depends upon the fund from which accumulation can be made; and upon the strength of the motives which induce saving, But it does not follow that all accumulation is used as capital. All funds that are invested have been saved; but all funds that have been saved are not necessarily invested. The motives for investment are included in the motives for saving. But as they form a part only of the latter class, it follows that saving may, and it in fact does, take place in the absence of those motives which are common to it and to investment. The motives to save exist in full force independently of

---

* *Transactions of the Central Relief Committee of the Society of Friends during the Famine in Ireland*, pp. 220 and 256.
† Levi's *Annals of British Legislation*, vol. ii. p. 178.

all considerations as to the additions which may be made to the savings themselves. It is undoubtedly a strong additional inducement to save when it is known that these savings will be immediately productive of a large profit; and the force of this influence is shown in the great savings by which any demand for new capital, as for railways or for similar purposes, is quickly supplied without disturbing existing investments. But even without this inducement, the necessity of providing for the wants of sickness or of age, the desire of assisting children or other relatives, the influence the respect and the other immediate advantages which attend the possession of wealth, amply suffice to induce men whose intelligence or affections are even moderately developed, to sacrifice the present to the future. Even in those countries where the state of society is such as practically to prevent any investment, the practice of hoarding is extensively prevalent. In the disorderly times of feudal Europe, treasure trove, as Adam Smith remarks, formed no inconsiderable portion of the royal rights. So late as the reign of Charles II., men of high official position * buried their money in gardens or hid it in butter firkins in the cellars of the tower. In the earlier part of the reign of William III., the greatest writers on currency were of opinion that a very considerable mass of gold and silver was hidden in secret drawers and behind wainscots.† The people of Hindostan in like manner frequently hoard large sums, often in the shape of personal ornaments of silver. In almost all eastern countries indeed, the practice of hoarding is still very general. In Yemen for example, and the countries bordering the Red Sea, it has been esti-

---

* See *Pepys's Diary*.
† Macaulay's *History of England*, vol. iv. p. 820.

mated that about one and a quarter millions of dollars find their way in the ordinary course of the Hegaz trade from Suez to Geddah; but none of the silver thus remitted ever returns. After the requirements of the local currency have been supplied, every proprietor conceals his accumulation in some secret treasure chamber, which with his own hands, and in the strictest privacy, he constructs for this purpose. Treasure trove is consequently very common in those countries; and the secret hoards sometimes amount, it is said, to 400,000 dollars. The great absorption thus caused may be conceived, if we suppose a population of 100,000 heads of families to hoard as an average 200 dollars each. Twenty millions of dollars, or upwards of five millions of pounds sterling, are thus left idle.*

But although saving differs from the productive employment of that which is saved, it is manifest that the latter is supplied from, and that its amount is limited by, the former. The hoard may be, and usually is, converted into capital. We have therefore to consider the circumstances in which the conversion takes place, or in other words the conditions under which men will consent to invest their savings. These conditions it is not difficult to ascertain. Since it is in expectation of certain advantages that the saving has been made, the inducement to expend the amount saved must be the hope of some additional advantage. This additional advantage is the prospect of gain, that is of increasing the amount of the sum already saved. Men must therefore believe that from the investment of their savings some gain, sufficient in their estimation to compensate their self-denial will accrue. They must further believe that this gain will accrue to themselves and

---

* Dassy on the *Commerce of the Red Sea*, quoted in *Journal of Statistical Society*, vol. xxiii. p. 472.

not to others; that is to say, that they will themselves enjoy the fruits of their industry. Unless such an advantage actually exist, and unless a man can realize it, his efforts to attain it, and the capital by which he assists them are alike thrown away.

§ 7. The first condition of investment is that the labourer (assuming him to be conscious that an advantage exists as the recompense of his forbearance) shall be satisfied as to the adequacy of this advantage. The question as to the adequacy of the advantage depends partly upon the magnitude of the assistance that capital in each case renders to labour; and partly upon the desires of the labourer. The inducement to accumulate is great or small when there is a great or a small return to labour assisted by capital, that is by the result of a given amount of self-denial during a given time. The sufficiency of the return where the amount and the duration of the forbearance is the same, varies in different circumstances. Where the risk is great, or where forethought and self-control are not habitual to the labourer, the return must be considerable. Where, on the other hand, there is little or no uncertainty, and where forbearance involves no serious effort, a small return will suffice. But as there is always an effort, there must always be a return for that effort. Accordingly, there is at every place and in every time a certain minimum rate of return below which the labourer will not then and there accumulate capital. As this limit is approached, the inducement to accumulate diminishes. When it has been attained, no further increase of capital can take place unless and until industry find some new direction or some superior subject matter.

§ 8. The next condition of investment is that the owner shall have a reasonable probability of enjoying the advantage derived from his forbearance. I have already observed that such a sense of security is essential to energy of labour; and the reasons for its importance in regard to labour equally apply in the case of capital. As labour for an immediate object implies an effort, so forbearance in using the results of such labour implies a further effort. If the motives be taken away, neither of these efforts will be made. When therefore this probability of enjoyment is from whatever cause diminished, and especially when the diminution arises from circumstances beyond the owner's control, the tendency to accumulate is checked. Hence, where the prospects of life are precarious, where the climate is unhealthy, or where the occupation is dangerous, or where an epidemic disease is prevalent, men are not inclined to save. It is said that the residents in New Orleans, the West Indian islands, and India are profuse in their expenditure; but when they remove to any of the European capitals and are not drawn into the extravagances of fashionable life, they gradually become as economical as other people. It has also been remarked that sailors and soldiers are usually extravagant; but this extravagance, so far as it really exists, may perhaps be ascribed not so much to the hazardous nature of their occupation as to the weakening effects upon their habits of self-reliance and self-control of the discipline to which they are subjected, and their isolation from the ordinary affairs of life. In later times improvements in the general intelligence of these classes, and the increased facilities for saving that have been afforded them, have greatly diminished their habits of reckless expenditure. The charge of profusion seems to be brought with greater reason against the dry grinders of Sheffield. These men, whose average duration of life is very short, are said to have refused to wear the magnetized masks that were contrived to arrest the steel filings from their work, and so to secure them against their peculiar pulmonary disease, thus deliberately preferring their present high rate of wages to a longer period of life. In like manner it has been universally observed from the time of the great Athenian historian downwards, that in seasons of plague men frequently rush into the most lavish profusion. A sort of despair seems to seize the mind; and a narrow selfishness, regardless of all save the individual and the present moment, is engendered.

In a country such as England, where the rights of property are ascertained and respected, and where the opportunities for investment are numerous, one of the principal causes of this feeling of insecurity is the instability of banking institutions or other companies to whom money has been confided. "The practical discouragements," says Mr. Greg,* "which have resulted from the absence of this due provision [*i.e.* for the free development and adequate security of saving] can be appreciated only by those who have come into close contact with the operative poor. Every defaulting savings bank, every absconding treasurer to a sick club or a friendly society, every bankrupt railway, every fraudulent or clumsy building league, every chimerical or mismanaged land association, preaches a sermon on the folly of frugality and providence not soon forgotten and not easily counteracted." Men feel that without any fault of theirs, without the occurrence of any circumstances which they could control, the results of their self-denial are entirely lost; that the good which they had anticipated has not been obtained; that the good which they had it in their power to enjoy has passed away; and they say in the melancholy words of the wise king of old, "The wise man's eyes are in his head, and the fool walketh in darkness, yet one event happeneth to them all."

---

* *Essays*, i. 394.

Another powerful cause of this feeling of insecurity, which, although I somewhat anticipate, may still be mentioned here, is the sense of legal oppression. This may arise either from direct tyranny on the part of the government, or from some interference of the law in favour of some other person against the rights of the person accumulating. Of the former, most Asiatic governments are conspicuous examples. Of the latter we have every degree, from actual slavery down to tenancy at will. No man will voluntarily expend money on the property and for the benefit of another. If he have no other mode of investment except upon such property, a man may indeed make accumulations, but he will never invest them. Similar consequences, although in a less degree, follow from that unsettled state of feeling which arises from any serious political agitation. In the days of the Revolution the English Government could not borrow at a lower rate of interest than ten per cent. During the agitation that prevailed in Ireland for Catholic Emancipation and subsequently for the Repeal of the Union, English capitalists were very reluctant to make advances upon Irish securities. A singular proof of this reluctance is found in the precedents commonly used in conveyancing. In order to facilitate the introduction of English capital into Ireland, a statute was passed authorising trustees to lend trust-money on Irish securities, unless expressly prohibited by their grantors. To avoid the operation of this Act, it has become the practice in all the common forms of English settlements and wills to add, as of course, where the usual power of investment is given to trustees the ominous words "but not in Ireland." This feeling of uncertainty also arises where the political agitation is not of a revolutionary character, but relates only to the continuance of the national policy on some particular question. While it is doubtful how long the existing arrangements may continue, or to what extent they may be changed, men although they will not cease to accumulate will abstain from investing their accumulations in that branch of industry which is the subject of dispute.

§ 9. "If consumption," says Mr. Malthus, "exceed production, the capital of the country must be diminished, and its wealth must be gradually destroyed from its want of power to produce : if production be in a great excess above consumption, the motive to accumulate and produce must cease from the want of will to consume. The two extremes are obvious : and it follows that there must be some intermediate point, though the resources of political economy may not be able to ascertain it, when, taking into consideration both the power to produce and the will to consume, the encouragement to the increase of wealth is the greatest." We may now perhaps attempt to ascertain this intermediate point with somewhat more precision than without keeping in view the true function of capital would be practicable. Since capital is an aid to labour, its accumulation will never exceed that point at which in each particular case it aids labour. When it ceases to render such aid, its accumulation becomes burdensome and is consequently checked. But up to that point, if circumstances be favourable, it will continue. The proof of these propositions may be readily stated. The accumulation of capital implies forbearance. Forbearance implies a painful effort. No such effort is voluntarily made except under the influence of some counteracting motive. The motive which leads to the accumulation of capital is the assistance which it affords to labour. As long therefore as that motive is in operation, as long as capital gives increased efficiency to labour, but no longer, accumulation may continue. When that motive ceases to operate, when there is nothing to compensate the effect of forbearance, accumulation will cease. If the circumstances be sought in which capital ceases to increase the efficiency of labour, the answer is that the increase of capital is no longer serviceable when the labourer possesses sufficient capital to enable him to devote to his work (whether he works alone or has the control of other labourers, and whether he consumes the produce of his labour directly or by exchange) so much of his time and power as he is able or willing to spend in labour; and also to enable him to turn to the utmost advantage, according to his knowledge, the natural agents at his disposal. Capital sufficient for these purposes or any less amount he can acquire. More than this he will not intentionally keep.

Thus then while it is true that as it has often been observed industry is limited by capital, it is equally true that capital is limited by industry. Capital is so potent an ally that no appreciable advantage in the war against the wilderness can be obtained without its presence; but it is only an ally, and requires the impulse and direction of the person whom it assists. Or to vary the metaphor, it is a force which requires some space for its operation. It must have what has been called a field of employment. Upon the extent of this field the amount of capital depends. In favourable circumstances it will gradually expand so as to occupy the entire field; but its operation is confined within that limit. This restriction applies only to capital properly so called. There may be an amount of accumulated wealth greater than the field of employment will bear. The motives to save continue to act even though investment be impossible. Men desire to provide for sickness or old age, to educate or advance their children, to promote the interests of other relations and friends, to assist objects of public or private utility, or to retain their superfluous wealth until some better opportunity for spending it to their satisfaction presents itself. The wealth so saved is rather potential than actual capital. It does not immediately aid labour, although it, or some of it, may at some future time be applied to that purpose. It cannot therefore be at the time of saving regarded as capital, although it may subsequently become so. It will become so if the field of employment be enlarged, that is if an opportunity be afforded for advantageous investment. The extent then to which it can be immediately applied in the assistance of labour is the limit of capital in the strict sense of that term. The extent to which there is reason to believe that it may hereafter be so applied is the limit of potential capital, or of capital seeking investment. It is obvious that the latter can be converted into the former by enlarging the field of employment, or in other words by enabling more labour to be employed, or the same labour to be more advantageously employed than before. Such an enlargement renders productive resources previously idle or only partially developed. It converts into an instrument for procuring enjoyment labour which before failed to a greater or less degree to effect any such purpose. But the capital of itself produces nothing. It only maintains and assists the labourer. Labour and the objects upon which it is applied constitute the field for the employment of capital; and capital finds in this field the maximum limit of its application, just as it had previously found in the productivity of labour the maximum limit of its existence.

CHAPTER X.

OF INVENTION.

§ 1. Invention, considered as an aid to industry, consists in the addition to human exertions or in the substitution for them of brute or inanimate forces. This contrived and as it were additional operation of natural forces must be distinguished from the ordinary co-operation of nature with man in production. Man can create nothing. He can merely produce changes in the existences around him. There is consequently no effort of his the success of which does not depend on external influences. His sole duty is to effect the appropriate changes; to place natural objects in the position proper for his purposes. The forces of nature do of themselves all the rest. But in addition to this co-operation of natural forces and human efforts, it is frequently possible to substitute in such co-operation for the efforts of man or to add to them the action of some other natural force. Some natural forces, when certain changes have once for all been effected in them, are capable of continually placing other natural forces in that position in which their results will be useful to man. Nature thus not only performs her own share of the work of production, but also does man's work for him; it is the latter operation that invention considered as an auxiliary to industry denotes.

It is also important to distinguish between invention and the agencies through which it operates. The latter may be comprised under the general name of implements, and form a part of capital. They are prepared for an object more or less remote by a greater or less exercise of forbearance. They are commodities that have been saved, and are applied to reproductive purposes. But they are usually the agents through which the natural forces work; and are not the natural forces themselves or the results that these forces accomplish. A steam-engine is included in the capital of its owner; but the invention consists in the application to human purposes of the elasticity of steam. The engine supplies the necessary conditions for the exercise in the required direction of the natural force; but the invention, the auxiliary to industry, is the power of that natural force to produce certain results which man, if he desired to obtain them, must otherwise have produced, either by some other natural agent or ultimately by his own muscular powers. There are instances in which the assistance of invention may be obtained without the aid of capital. The action of the sun and air in bleaching, the motive power of water and of wind, the use of water in irrigation, the use of fire in clearing land, are examples of physical forces added to or substituted for human forces without the intervention of any other agency. The two ideas therefore, the invention, and the agency through which the invention usually operates, must be kept distinct. But whether invention acts directly, or as more frequently is the case, under certain costly conditions, its character remains the same. Man and natural agents together produce certain commodities. Man saves for future use some of these commodities under the name of capital. He perceives that he can make other natural forces do part of his work. To obtain the aid of these forces he often although not always employs, whether in the form of food, or of materials, or of implements, the capital he has accumulated. Thus invention with or without the aid of capital enables certain natural forces under man's direction to co-operate in his stead with those other natural agents which formed the subject matter of his exertions.

§ 2. It is not necessary in the present generation to pronounce any eulogium upon the advantages of implements. Amongst us such a discourse would be as superfluous as the panegyric of Hercules was amongst the Athenians. There never has been any difference of opinion as to the benefits arising from the use of these smaller implements generally described as tools; although the conviction of the good resulting from machinery has not been held with equal unanimity. Yet even the doubts that lingered respecting the latter subject have, except where personal interest has biassed the judgment, melted away before the great industrial successes of the present reigns. Nothing now remains to be said upon this topic, save briefly to illustrate the extent and to indicate the character of those advantages to which so many of our daily comforts are due.

M. Chevalier[*] has attempted to obtain numerical expressions for the present advance of productive power in the leading branches of industry as compared with the state of the same occupations at an earlier period. Such expressions can at best be very rough approximations. The data on which M. Chevalier founds his estimate are as regards the earlier period almost arbitrary ; and the industry of the last twenty years has taken away all claims to represent modern

---

[*] *Cours d'Économie Politique*, tome i. Leç. 2.

productive power from figures calculated in 1841. Still these calculations serve to give some general idea, and nothing more is required, of the extent to which we are indebted to invention. M. Chevalier estimates that, as between the labour of one man at the several periods mentioned and the labour of one man with all the industrial appliances available when he wrote, the increase of productive power in the manufacture of iron during the last four or five centuries has been as 25 or 30 to 1. In the manufacture of bread the increase since the time in which the Odyssee was written is 144 to 1. In cotton fabrics, within the last century only, the increase is as 320 to 1. The labour of one American in the transport of goods was in 1841 as effective as the combined labour of 6,657 subjects of Montezuma.

Several other attempts have been made to convey an adequate notion of the assistance thus rendered to man. In India for example a good spinner will not be able to complete a hank in one day. In Great Britain and America one man usually attends a mule containing 1,088 spindles. Each of these spindles spins in the day three hanks. Thus the assisted labour of the Anglo Saxon is to the unassisted labour of the Indian as 3,264 to 1. Even as between persons of the same nation, the assistance of proper machinery makes a very great difference. Forty years ago it is said that three men with the appliances then used could with difficulty make in the day about 4,000 sheets of paper. With the improved machinery now in use, the same number of men can produce daily 60,000.* Lace of an ordinary figured pattern used to be made on the cushion by hand at the rate of about three meshes a minute. In Nottingham a machine attended by one person now produces lace of a

---

* *Waste Products*, p. 133.

similar kind at the rate of about 24,000 meshes a minute. Thus the proportion of the produce of assisted to that of unassisted labour is as 8,000 to 1.* Again, it has been estimated that three tons of coal give an amount of force equal to that of an adult man working during twenty years 300 days in each year. It is assumed that of the total amount of coal raised in England one-sixth part is applied to the production of force. Hence it follows that England has an army of nearly three and a half millions of men labouring in her behalf during the period above mentioned without any charge for their maintenance. It is however a very different thing to obtain the same result by the mere labour of a sufficient number of men and by one man with the assistance of proper implements. Even if the immediate material result be in both cases equal, the expense at which it is procured, and that too reckoned in something more important than money, is very different.

The recent progress of agriculture also presents some striking evidence of the influence of invention. This case furnishes a good example of invention as distinguished from either discovery or improvement, since agricultural implements "are intended not to bring about new conditions of soil, nor to yield new products of any kind;† but to do with more certainty and cheapness what had been done hitherto by employing the rude implements of former centuries." Mr. Pusey,‡ in his Report on Agricultural Implements in the Exhibition of 1851, thus sums up the advantages derived within the twelve years next preceding his Report from the extension to agriculture of mechanical inventions. "To ascertain the amount of saving precisely is difficult; but looking through the successive stages of management,

---

* Whitworth: *Special Report on New York Industrial Exhibition*, p. 42.
† *Reports of Juries; Exhibition of* 1851, p. 225.
‡ Ib., p. 240.

and seeing that the owner of a stock farm in the preparation of his land by using lighter ploughs is able to cast off one horse in three; and by adopting other simple tools to dispense altogether with the great part of his ploughing: that in the culture of crops by the various drills horse-labour can be partly reduced; the seed otherwise wanted partly saved; and the use of manures greatly economized; while the horse-hoe replaces the hoe at one-half the expense: that at harvest the American reaper can effect nearly thirty men's work; while the Scotch cart replaces the old English waggon with exactly half the number of horses; that in preparing corn for man's food the steam threshing-machine saves two-thirds of our former expense; and in preparing food for stock the turnip-cutter, at an outlay of 1s., adds 8s. a-head in one winter to the value of sheep: lastly, that in the indispensable but costly operation of drainage, the materials have been reduced from 80s. to 15s., to one-fifth namely of their former cost: it seems to be proved that the efforts of agricultural mechanists have been so far successful as in all these main branches of farming labour taken together to effect a saving on outgoings or else an increase of incomings of not less than one-half."

§ 3. The principal advantages which the use of implements secures to man are the increase to his powers or the saving of his toil; the saving of his time; the greater certainty of his work; the economy of his existing materials; and the power of using superior materials. As to the first of these advantages, the fact and the gain that results from it are sufficiently plain. The increase of power and the saving of exertion are the primary purposes for which implements are used. There are many desirable objects which the human strength is quite unequal to accomplish. There are many others which it could accomplish, but only by a great expenditure of labour. These objects the proper use of natural forces can accomplish, and that too at a comparatively trifling cost. The removal of immense masses of matter, such as large trees or stones or similar objects, illustrates these advantages. By one man they cannot be even stirred; the combined efforts of many men suffice in a long time and at a great cost to remove them; by proper machinery the same result, otherwise so tedious and so costly is quickly and cheaply obtained. By the consumption of a bushel of coals in the furnace of a steam-boiler a power is produced * which in a few minutes raises 20,000 gallons of water from a depth of 350 feet. The same result would require the continuous labour of twenty men during an entire day with the common pump. It would take a single workman many years, if indeed he could at all accomplish the task, to raise so much water with a bucket and rope; and without these contrivances, themselves no unimportant invention, the difficulty would far exceed his unaided powers.

There are two modes in which these implements operate: one is by giving increased mechanical advantage to the exercise of muscular power; the other is by substituting for that power some more potent natural agency. Thus it appears as the result of careful experiments that the force necessary to move a stone along the roughly chiselled floor of its quarry is nearly two-thirds of its weight; to move it along a wooden floor, three-fifths; by wood upon wood five-ninths; if the wooden surfaces are soaped, one-sixth, if rollers are used on the floor of the quarry, it requires one thirty-second part of the weight, if they roll over wood, one-fortieth, and if they roll between wood, one-fiftieth of its weight. "At each increase of knowledge," says Mr. Babbage,† " as well as on the contri-

---

* Porter's *Progress of the Nation*, p. 274.
† *Machinery and Manufactures*, p. 7.

vance of every new tool, human labour becomes abridged. The man who contrived rollers invented a tool by which his power was quintupled. The workman who first suggested the employment of soap or grease was immediately enabled to move without exerting a greater effort more than three times the weight he could before." In these experiments the muscular powers of the labourer was exerted with at each step of the process a constantly increasing mechanical advantage. But it is only necessary in such cases that some force should be used, and not that the force so used should be the contraction of the human muscle. It may be the power of a horse or of wind or of steam. Man may therefore substitute for his own force these or any other suitable powers. The amount of this substitute is determined by the same principle as that which determines in any case the quantity of human effort, the degree, namely, of mechanical advantage with which the force is applied.

§ 4. The example already cited of a steam-engine raising water also illustrates the distinction between mechanical advantage and motive power. No man by the mere effort of his naked hands could displace so much water so circumstanced. With the help of a rope and a bucket the undertaking becomes at all events possible. If a windlass were added, the muscular power of the labourer would be exercised with much greater advantage than before. If this apparatus were changed for the pump, his power would be exerted in still more favourable conditions. But the steam-engine introduces a new and greater force to replace the action of the muscles. Besides the advantage thus gained in labour the new force also effects a great saving of time. The cost even with the pump could not be less than 50s. With the steam-engine it is only a few pence : the pump required a whole day ; the engine does the same work in a quarter of an hour. There are many examples too where the whole benefit consists in the saving of time. In the Mint of the United States, whence seventy millions of coins are issued within the year, two female manipulators checked by a weighing clerk have been able to count these enormous numbers. A counting frame or tray has been contrived, by the help of which in a few seconds 1000 coins can be exactly measured. It is said that the same labour could not be performed in the same time by less than thirty or forty persons; while from the strain upon the attention incident to such an operation equal accuracy could not be attained.*

Another illustration of the influence of invention in economizing time is found in the case of a daily paper. Time is manifestly of the essence of journalism. Under pain of the great depreciation, if not of the absolute loss, of its value, a daily paper must be published not only on each day but at a given hour each day. No number of skilled copyists could supply the present demand for the London *Times*. If printing and printing-machines had not been invented, no such demand could have grown into existence. Another interesting example is found in the case of bleaching. In the last century our linen was sent to Holland to be bleached. "The Dutch," says Mr. Knight,† "steeped the bundles of cloth in ley made by water poured upon wood ashes, then soaked them in buttermilk, and finally spread them upon the grass for several months. These were all natural agencies which discharged the coloring matter without any chemical science. It was at length found out that sulphuric acid would do the same work in one day which the buttermilk did in six weeks, but the sun and air had still to be the chief bleaching powers. A French chemist then found out that a new

---

* Professor Wilson's *Special Report on New York Exhibition*, p. 59.
† *Knowledge is Power*, p. 235.

gas, chlorine, would supersede the necessity for spreading out the linen for several months; and so the acres of bleaching which we were using in England and Scotland, for we had left off sending the brown and yellow cloth to Holland, were free for cultivation. But the chlorine was poisonous to the workmen, and imparted a filthy odour to the cloth. Chemistry again went to work; and finally obtained the chloride of lime, which is the universal bleaching powder of modern manufactures. What used to be the work of eight months is now accomplished in an hour or two; and so a bag of dingy raw cotton may be in New York on the first day of the month, and be converted into the whitest calico before the month is at an end."

§ 5. Another advantage of invention is the certainty which in many cases it is able to ensure. Many operations, although time is otherwise of no peculiar importance in them, can only be performed in a particular state of weather. As the weather is liable to constant and unforeseen changes, these operations become uncertain. It is therefore important either that they should be rendered independent of atmospheric changes; or that they should be conducted with such expedition as to be completed within the usual intervals of propitious seasons. Both these results are obtained by the aid of invention. The steadiness with which steam did its work in all seasons and in all circumstances was at once felt, in all cases where the two classes of agents entered into competition, to be one of the chief advantages it possessed over the power obtained from wind or water. Our navigation is now no longer dependent upon the accident of the rising or the direction of a breeze. Our mills no longer dread the consequences of a drought. In 'agriculture, by the use of proper machinery water can be applied at the place and at the time at which it is required, or withdrawn with almost equal facility. In a late season the safety or the loss of a whole crop may be determined by the use of the reaping machine. When the harvest is gathered, steam threshing-machines will, within the limits of a few fine hours, thresh in the open air all the corn, and thus dispense with the necessity of extensive and costly barns. "The main difficulty of farming," says Mr. Pusey,* "has always lain in its uncertainty. Though machinery has not altogether cured, it certainly has much mitigated this evil. On undrained clays, a wet winter may destroy half the yield of the wheat. On the same land drained, the wheat may escape altogether unhurt; and you may also plough heavy land in wet weather when drained although you could not before. Upon any land wheat may suffer in winter, but in spring the presser settles it in its bed, and the manure-distributor with a cheap sprinkling restores it to vigour. In sowing barley, earliness may save the crop, but the ground is often too cloddy, though the season is wearing away and May drought approaching. This cloddiness may be prevented by the paring plough, or if it could not be prevented, may be remedied by the clod-crusher or the Norwegian harrow, and besides these implements the cultivator does the plough's work in one-fourth of the former time, thus enabling the farmer to profit by the auspicious hour of seed-time. And so too with the turnip. The land being prepared for it in the previous autumn and winter, is moist to receive the seed: the dry-drill supplying it with super-phosphate saves it almost certainly from the fly, or yet more the water-drill, anticipating the clouds, makes its seed-time independent of weather, while the horse-hoe afterwards preserves it from neglect in the busiest harvest-time. Again, while machinery remedies the absence it also guards against the inconvenient arrival

---

* *Reports of Juries*, p. 241.

of rain, by making hay and now even reaping corn while the sun shines. It may be further said then that machinery has given to farming what it most wanted, not absolute indeed but comparative certainty."

§ 6. There are numerous economic advantages attendant upon invention. It can convert mere refuse into valuable commodities. It can turn the filthy and worthless rag or the sweepings of the cotton-mill into fine and beautiful paper. It can produce from sand and soda brilliant glass. It can obtain from kelp a number of useful and expensive substances. It can produce from sawdust nutritious bread; and the choicest perfumes from the vilest refuse. Frequently, too, substances are thrown away as worthless long before their useful properties have been exhausted. Quantities of washdirt have been abandoned on the gold-fields which with proper treatment have afterwards yielded an adequate remuneration. The mines of Reichenstein in Silesia which were abandoned for five centuries have been recently opened with advantage in consequence of the application on a large scale of a method invented by Professor Plattner for separating gold from the waste of arsenical ores. The same principle is equally applicable to the vast quantities of refuse that have accumulated near other old works.* Again, most of the ores of lead contain a small portion of silver. The usual mode of obtaining this silver was so costly that it could only be applied with advantage when the lead contained at least twenty ounces of silver to the ton. A new method, founded on the property of bodies to separate during crystallization, was introduced by Mr. Pattinson, under which lead containing only three ounces of silver to the ton may be profitably treated. The consequence has

---

* *Reports of Juries*, p. 5.

been that several lead mines have been worked with profit which would otherwise have been neglected.* So, too a great impulse has been given of late years to the woollen manufacture in Yorkshire by the introduction of the process termed rag-grinding. Old woollen rags are torn up by proper machinery; and after having been thus reduced to their original state of wool are subsequently retransformed in the usual manner into cloth. The cloth which, according as it is made from rags of soft materials or of woollen cloth, is called shoddy or mungo, is from the extreme shortness of the fibre weak and tender. There is however a great demand for goods in which substance and warmth are the chief requisites, and in which strength is of comparatively little importance. Such, amongst others, are paddings, linings, the cloth used for rough and loose great coats, for office coats, and even for ladies' capes and mantles. For all these purposes this inferior cloth is perfectly suitable, while its price is considerably less than that of broadcloth.† Where therefore this cloth is fairly sold for what it really is, and where its character is not fraudulently misrepresented as it too often has been, the utilization of these rags is a legitimate and important improvement in manufacture. Of such consequence to the trade is this process that in the town of Leeds alone it adds to the annual stock of wool an amount equal to the fleeces of 400,000 sheep.‡

The commercial success of many manufactures depends upon their power of turning to profitable account their refuse materials. This kind of economy has been carried to a great length in the uses to which the horns of cattle are applied. They are first sold with the skin to the tanner. By him they are sold to

---

* *Reports of Juries*, p. 4.
† *Journal of Statistical Society*, vol. xxii. p. 30.
‡ Ib., vol. xxi. p. 435.

comb-makers. After various processes, one portion of the hard part of the horn is made into combs, another into the commoner kind of lanterns, and a third into the handles of knives, the tops of whips, and similar objects. From the softer part of the horn when boiled down, there is produced a quantity of fat that is used for making yellow soap; the liquid is used by cloth-dressers for stiffening, and the residuum of bony substance is ground down for manure. The clippings of the comb-maker and the shavings of the lantern-maker are also used for manure. The latter are sometimes made into toys from their property of curling up when placed in a warm hand. In France this utilization of worthless animal substances has been still further extended. In the Exhibition of 1851* special mention was made of the beautiful and diversified products fabricated from such substances. They include different kinds of gelatine in thin layers adapted for the dressing of stuffs, and for gelatinous baths in the clarification of wines which contain a sufficient quantity of tannin to precipitate the gelatine; pure and white gelatines cut into threads for the use of the confectioner; very thin white and transparent sheets called papier glacé or ice paper for copying drawings; and, finally, a quantity of objects of luxury or ornaments formed of dyed, silvered, or gilt, gelatines adapted to a variety of purposes, and to the fabrication of artificial or fancy flowers.

Invention also finds means not only of converting the worthless into the valuable, but of separating and disincumbering the useful from that part of it which is useless. This advantage is felt most strongly in relation to transport. In cotton a large portion of the crop when gathered is refuse. Accordingly in all places where the growth of cotton is an established branch of industry, it is cleaned by the gin, a

---

* *Reports of Juries*, p. 70.

machine invented for that purpose. In the same manner wool in the grease as it is termed is very much heavier than when it has been freed from its impurities. Care is therefore taken to cleanse it by water-power or otherwise before it is shipped. If the cotton and the wool were exported without any preliminary preparation, the bulk and weight of the bales would be greatly increased, and it would be necessary to pay, as the Chinese in their trade with Java actually do,* a large sum for transporting from America or Australia to England a quantity of dirt. In the case of Peruvian Bark† it was discovered that the medicinal qualities for which it was esteemed resided not in the bark but in the alkali (quinine) which the bark contained. About one ton of this substance can be obtained from forty tons of bark; and accordingly for every such ton thirty-nine tons of rubbish are conveyed across the Atlantic. Nothing but the disorder and poverty of South America can account for the medicine not being extracted on the spot, and for so much unnecessary outlay being incurred.

§ 7. Invention can thus utilize substances which at first sight appear wholly worthless; or which when worked by unassisted labour are at least unremunerative. But in addition to this power, it also enables the labourer to work materials which, although in themselves amply remunerative, were previously beyond his reach. The more useful an agent is when under proper direction, the more difficult is its conversion to human purposes. The forces in which its utility consists, spontaneously exert themselves in a manner that is not adapted to the wants of the labourer; and present, before their direction can be changed, a formidable resistance. Frequently too these useful substances are

---

* See Babbage, p. 19. † Ib., p. 314.

difficult of access or of removal; or are combined with other materials which interfere with their use, and from which they cannot easily be severed. Some of the most fertile mines are hundreds of feet below the surface of the earth; and the character of the strata through which the miner must pierce is such that there is a constant accumulation of water. It becomes impossible therefore to reach the material without the aid of elaborate and costly implements. The ground is pierced, the rocks blasted, the rubbish removed, the water pumped out, the mines ventilated, the workmen raised and lowered, the mineral extracted and brought to the surface, not by the mere action of human muscles, but by the aid of some mechanical contrivance or some additional motive power. If our supply of the most important minerals depended upon the quantity that we could pick up from the surface of the ground or that we could obtain by scraping a few inches or even feet below the surface, the whole condition of society would be very different from what it now is. The loss would be still greater, if we were further limited to those minerals only which are found pure. There are few metals that are found in a state fit for immediate use; and the quantities even of such metals are not great. But mere hand-labour could never separate, in some cases at all, in all cases to any considerable extent, the ore from the substances with which it is mixed. The union is either chemical or of a mechanical nature so intimate that hand-labour even assisted with tools would be altogether futile. Even therefore if the difficulty of extracting the ore was surmounted, its preparation would without the use of some other natural agent be impossible. It is to the facility that from the peculiar distribution of its natural agents it possesses of smelting its iron by means of its coal, and of extracting its coal by the use of its iron and by the assistance that the coal itself can render, and of transporting both metals cheaply and expeditiously wherever they are wanted, that the manufacturing pre-eminence of England is in great measure due. The ironstone and the coal are in many districts found in close proximity. There is in such cases a substratum of limestone which serves as a flux for melting the metal. Beds of clay are generally found in these formations, and furnish valuable materials for the construction of furnaces. The insular position of the country and its extensive means of inland navigation admit of the coal or the ore or their products being readily removed from place to place. And the coal supplies not only the fuel that is required for acting upon the iron, but the motive power by which it is itself extracted from mines otherwise inaccessible, and by which the iron when prepared is made to assume the desired form.

§ 8. There is another effect of invention which must not be overlooked. We have seen the extent to which it saves the labour and the time of the workman. That labour therefore and that time are free to be applied to other industrial purposes. It is a manifest waste of power if a man himself perform labour which could be done equally well by some inferior agent; and the gain therefore of placing man in his fit position, that of superintendence, in relation to the other forces of nature, is very great. But this advantage is not the only one. The saving of labour and of time not only admits of an improved industrial organization; but it represents the relief from sordid cares and exhausting toils, and the acquisition of a leisure which may be applied to worthy objects. Men must indeed live, and work to secure the means of their subsistence; but in the present day no less than in the time of Juvenal it is true that for the sake of life they often lose the motives for living. The man who passes his whole existence in a prolonged and hopeless struggle against nature is rather a slave than a freeman. When he has conquered nature by obeying her, he assumes his rightful position of master upon the earth, and all the forces of nature do his work. The great results of invention are not only increased products, but greater facilities for enjoyment, and decreased risk and decreased suffering. An Eastern tyrant is said to have set the whole male population of his province to clear out with their unassisted hands the mud which had choked up a disused canal. Tens of thousands died in the tedious execution of a work which with proper machinery could have been promptly and safely performed. The boatmen of the Rhone, the carriers of the Andes, the workers in phosphorus, the pressmen before the use of steam-printing, miners of every kind before the invention of the Davy lamp, and many other classes of workmen were subject to daily risk or to grievous diseases. These risks and these diseases have been removed by the aid of inventions; and the quantity of wages earned has not decreased. In many of these cases the depression produced by their employment almost necessitated the fatal stimulus of drink; and thus the lives of the workmen, short as they were, were still further shortened and embittered by intemperance. The tendency of the extension of machinery is to throw upon machines for the benefit of the labourer all the most dangerous and the most degrading occupations. The men therefore who have been hitherto engaged in these occupations will become more respectable, more temperate, more healthly, and consequently more rich. It is by no means indifferent to society, even supposing that the cost were in both cases equal, whether the same quantity of goods be transported upon men's backs or by a steam-engine. "What comparison," says M. Chevalier,* "can one make between the condition of

---

* Tome i. p. 325.

a wretch plodding under his burden, and of the mechanist who directs and controls at his will his admirable engine. The one is degraded under the weight of his task: the other accomplishes a work of which he may be justly proud; and where it is his intelligence much more than his muscular force that is called into play."

CHAPTER XI.

OF THE CIRCUMSTANCES WHICH DETERMINE THE EXTENT OF INVENTION.

§ 1. Invention denotes the substitution in industrial operations of natural forces for those of man. Since these natural forces are directed to a given end, the person who so directs them must be aware of their existence; and must be more or less familiar with their properties and relations. But as natural forces operate only under precise conditions, something more is required than an acquaintance with the character of these conditions. There must exist the means of practically fulfilling them. It is often difficult to bring together in a convenient form or at the requisite time the necessary assemblage of these conditions. Considerable ingenuity is required to effect this purpose, and to avoid or counteract the operation of disturbing forces. In accomplishing these objects, very delicate manipulation is frequently essential. The more powerful the force which is subdued to human purposes, the more elaborate is the workmanship of the machinery through which it acts. Thus in addition to the theoretical knowledge, invention demands manual dexterity, or at least the capacity of acquiring it. The machinery for the industrial operation of natural forces must obviously have a tangible form, and be constructed out of other substances. Consequently invention is limited by the commodities which must be used in preparing the framework through which the natural force has to operate. Each of these depends in its turn upon similar conditions; and thus the whole circle of the arts is concerned in each individual invention. We thus find a further limitation for invention in the existing state of the constructive arts.

Invention proceeds from a certain motive. Its extent, or at all events the attempts to increase its extent, will be determined by the intensity of that motive. The motive of invention is the assistance that it renders to industry. It is therefore likely to be most sought when industry most requires assistance. Accordingly the desire for inventions is of hardly less importance in their development than the power to execute them. Thus then the extention of inventions depends partly on men's power and partly on their will. Their inventive power depends in the first instance upon the state of the physical sciences, upon the practical ingenuity available for carrying into effect the results of that science, and upon the dexterity that can be applied to the manual processes thus involved. A secondary class of limitations consists in the state of the kindred or subsidiary arts; and in the intensity of the demand for the aid that invention renders. This demand springs from the perception of some want, and the belief that this want can be most efficiently satisfied by the operation of some physical agency. In the concurrence of all these conditions the circumstances are found which are most favourable to the increase of inventions.

§ 2. The systematic knowledge of the relations which exist between the forces of nature is of late growth. Its generalizations pre-suppose the existence of a great mass of experience from which they are obtained. But long before the theory of any subject is complete, this experience groups itself into maxims which are often of great practical utility. The knowledge that is imperfect for the purposes of speculation is often a sufficient guide for the daily business of life. Accordingly many of the most important arts have been practised, while the true principles on which they are founded were still undiscovered. Parchment and paper, printing and engraving, glass and steel, gunpowder, clocks, telescopes, the mariner's compass, were all known, and the processes which they severally imply were successfully practised, at a period when the corresponding sciences had no existence.* Even at the present day many processes of our most successful arts have not yet received a scientific explanation. In the manufactures of porcelain, of steel, of glass, and in many other arts, the chemical principles are yet unknown which would explain the conditions of success or failure in their manufacture.† But in all empirical arts, the limit of improvement is soon reached. The knowledge on which they depend is precarious; and is reliable only so far as the process and the materials are unchanged. It is only when the properties and the relations with which an art is concerned have been thoroughly investigated, and are accurately known, that that foresight can be obtained which leads to prompt and confident action. Science is the intermediate step between the empirical and the scientific state of art. It owes its origin to the one; it repays its filial debt by the generation of the other. Where therefore science is successfully cultivated, we may expect that if other conditions be fulfilled, inventions will be effective and abundant. That this explanation is not ill-

---

* Whewell's *History of Inductive Sciences*, vol. i. p. 252.
† Ib., 255.

grounded the numerous gifts which, especially within the present generation, science has bestowed upon the arts, sufficiently prove. Modern navigation is due to the speculations of the Alexandrian mathematicians. The electric telegraph is directly derived from the experiments of Œrsted, and the investigations of Ampere. Watt's greatest improvement in the steam-engine sprung from his steady apprehension of an atmological principle. The safety-lamp was the result of scientific inquiries by Davy. It is owing to the researches of chemists that a bale of cotton can be bleached within a few hours after it has been manufactured. Liebig has indicated to the farmers both the manures appropriate to each kind of soil, and the localities where the manures are likely to be found. Great as was the success of the practical builders of former days, nothing but accurate scientific knowledge could have enabled Stephenson to foresee that iron beams could be constructed 500 feet long, resting merely on their extremities, and capable of sustaining the weight of a railway train, and the jar of its unchecked motion. It may indeed be asserted that there is no extension, however apparently trifling, of our knowledge of nature which is not more or less fruitful in its applications to the uses of our daily life.

§ 3. There is little need in a country and an age of engineers and contractors to cite examples of the practical ability shown in the application to human purposes of advancing knowledge. The progress of science is now constantly and keenly watched for any practical advantage that it may afford to art; and every scientific discovery thus acquires almost immediately a commercial importance. We have examples of this connection in the eagerness with which Chemistry, Electricity, and Photography, have been respectively applied to the uses of life; and in the anxiety shown respecting the development of new qualities in gutta percha and similar substances. We are now so familiar with this kind of skill that examples illustrative of its absence will perhaps be more striking than those which describe its actual operation. It is well ascertained that the Chinese have been acquainted almost from time immemorial with the composition and the properties of gunpowder; yet they seem never to have devised the means of utilizing for military purposes this great force. Even when this knowledge had reached them from the West, their attempts to avail themselves of it were for a long time very imperfect. Their artillery, until the Jesuits taught them the art of casting cannon, consisted of iron tubes bound together by hoops. In the same manner, although they have long been acquainted with the properties of the compass, their knowledge of shipbuilding is so defective that they never venture upon any long voyage. In our country, Roger Bacon was acquainted with the theory of the telescope, yet it was many years after his death that the first telescope was constructed. The theory of the longitude and of the means by which it might be ascertained were long known: but the application of this knowledge to nautical purposes is comparatively recent. For upwards of two centuries the great sovereigns of Europe offered munificent rewards for the discovery of the means of ascertaining the longitude at sea. But it was only by the protracted labours of Harrison in the last century, and of succeeding artists who have improved his invention, that chronometers have been brought to their present perfection.

§ 4. But whatever may be the general state of scientific knowledge, or however great may be the ingenuity that an inventor possesses or can command, his labours are in vain unless a sufficient amount of manual dexterity be available for his purposes. If the workmen are tolerably intelligent, or have been accustomed to processes not very dissimilar to those for which their services are required, they can usually be quickly trained to the required degree of proficiency. This aptitude of the workmen is sometimes sufficient to determine the site of a new industry. It was the presence of the trained population in Manchester and in Paisley that drew the cotton trade to Liverpool and Glasgow rather than to the older and in other respects equally convenient port of Bristol. It was the facility of procuring workers from a population accustomed to the domestic manufacture by handicraft of lace that brought the lace trade, when driven by the suicidal violence of the working people from Nottingham, to its present seats at Tiverton, Barnstaple, Taunton, and Chard. But although after sufficient practice an inventor may train up a body of skilled workmen, yet in the earlier stages of almost all new inventions the want of such workers presents a formidable impediment. The alchemists were obliged to invent their own apparatus; to contrive the processes which served for the production of their preparations; and to make with their own hands everything that they used in their experiments.* The earlier printers made their own presses; cast their own types; set up their own copy; and struck off their own impressions. The Italian and the French engineers were at an early period of their art acquainted with the properties of iron, and frequently attempted to use it in bridge building. But the founders of that day were unable to cast iron in large masses: and the project was consequently abandoned.† It was with great difficulty that Watt could get the cylinders for his condensing engine cast straight. When Arnold the celebrated chronometer-maker, desired

---

* Liebig's *Letters on Chemistry*, p. 54.
† Smiles's *Lives of the Engineers*, vol. ii. p. 355.

to make for George III. a miniature repeating watch on a novel principle, he not only designed and executed the work himself, but had to manufacture the greater part of the tools that he required in its construction.* In former times in England foreign workmen were employed for almost every description of skilled labour; and even in the present unequalled state of the mechanical arts, inventors complain that the English artizan, perfect though he be in his usual routine, becomes as soon as he is placed in novel circumstances comparatively helpless.

§ 5. Some of the examples already cited for other purposes also serve to illustrate the proposition that no invention can be considered absolutely; but that its success has relation to the existing state of the arts. I will add a few additional instances. The art of navigation depends upon certain astronomical observations, and so upon the art of constructing astronomical instruments. The latter art in its turn depends upon the art of manufacturing glass; and the glass manufacture again involves several other distinct arts. To no science does industry owe more than to chemistry. But the success of modern chemistry is largely due to the superiority of the instruments it has been enabled to employ. "Without glass, cork, platinum, and caoutchouc," says Leibig,† "we should probably at this day have advanced only half as far as we have done. In the time of Lavoisier only a few, and those very rich persons, were able on account of the costliness of apparatus to make chemical researches." Again, the iron trade depends in great measure upon the use of coal in the furnaces, and this improvement, although actually introduced, remained unused for nearly a century,

---

* Timbs' *Stories of Inventors*, p. 119.
† *Letters on Chemistry*, p. 124.

until certain alterations in the blowing apparatus which were necessary for its proper effect were accomplished. The coal trade in its turn is dependent on various mechanical arts for the extraction of the mineral and its conveyance. Both cheap coal and cheap iron were essential to the use of gas as a means of illumination. Iron piping at a reasonable rate is also an essential condition of an abundant water supply. When Sir Hugh Middleton brought the New River to London, the most costly and most troublesome part of his work still lay before him. His only method of distributing the water to the various consumers was through wooden pipes; and it was not until a more efficient substitute became available that the system of water supply was generally established. The arts of inland and marine navigation, of engineering, of machine-making, the improved process of bleaching, and a thousand other arts, were all conditions precedent to the present development of the cotton manufacture. Again, photography requires the aid of iodine and bromine, of gun cotton and collodion. Besides the aid of the chemist, the photographer owes much to the skill of the optician and of the mechanist. So numerous indeed are the relations between the various forces of nature that it is difficult to mention one which is not directly or indirectly influenced by any improvement in the processes by which any other is worked.

§ 6. The principal circumstance, however, which affects the progress of inventions is the strength of the motive for their use. When the demand is sufficiently strong, the supply generally overtakes it. The old proverb that connects the way with the will finds in the case we are now considering abundant confirmation. Although the conditions previously specified must in every case be fulfilled, yet the strength of the desire generally leads sooner or later to their fulfilment. At every time and in every place the master of all arts and the bounteous bestower of genius is Want. It quickens the perceptions, and sharpens the power of contrivance. This action is felt in every part of industry; but is most perceptible perhaps in the subject-matter with which we are now concerned. Examples of its influence abound in every direction. There is scarcely one of the great improvements in the arts which has not immediately been connected with the expansion of some want. The want may be caused by the natural development of the capacities of the person by whom it is felt, or by the presence of some external obstacle. But whatever may be its origin, or whatever form it may assume, there is always a want, and a want of considerable persistence and intensity. Thus canals in England were made in consequence of the excessive cost of land transport between the seaports and the manufacturing towns. Sixty years after the opening of canals, the country had again outgrown its existing means of communication; and the interests of trade imperatively demanded some further improvement in locomotion. Of this want railways were the result. In the construction of these railways, it became necessary to carry the line over an arm of the sea in such a manner as not to impede the navigation. It was this want that led to the invention of the tubular girder and the tubular bridge. It was the increasing demand of an intelligent and interested population that led to the improvements which the printing press has received. It was the insufficient supply of cotton yarn in former days, and the annoyance and the waste of time to which hand-spinning gave rise, that led to the great inventions in which Arkwright led the way.* It was the gradual destruction of the old forests that led to the develop-

---

* *Pictorial History of England*, vol. vi. p. 489.

ment of the coal trade, and so indirectly to the still more marvellous expansion of the iron trade. In America where labour is dear and often inferior, many ingenious methods for enabling industrious people to dispense with such dubious help have been contrived. To this cause we owe amongst other improvements the cotton-gin, and the reaper, and the sewing-machine. It is well known that to the strikes of workmen, and the confusion thence resulting, many important improvements in machinery are due. Wars and hostile tariffs have a similar tendency. Some years since, the then Government of Naples attempted to establish a monopoly of the trade in sulphur. In the short period that that monopoly lasted, fifteen patents were granted for methods to recover the sulphuric acid used in making soda. Large quantities of this acid were also under the same stimulus obtained from iron pyrites; and if the advance in price had continued much longer, there is little doubt that an additional supply would have been obtained from gypsum.*
The exorbitant duties imposed by the Russian tariff have led English consumers to substitute the products of the palm tree and the cocoanut for the tallow oil and the linseed oil of Russia. To the great war of the First Napoleon and his commercial policy the development of the beet-root sugar trade is due; and the more important discovery of the production of soda from common salt. We have, too, before our eyes the eagerness with which in a great emergency substitutes have been sought, although ineffectually, for cotton.

§ 7. From these considerations some apparent peculiarities in the growth of inventions may be explained. It has often been remarked as a sort of mysterious coincidence

---

* Liebig's *Letters on Chemistry*, 152-3.

that important inventions and discoveries have been made almost simultaneously by independent inquirers. But there is nothing mysterious in this fact. There is a concurrence of the conditions under which inventions take place. Some advance in science, or some improvement in art is effected. A sufficiency of means to prosecute the particular line of inquiry is forthcoming; or the stimulus of some new want is felt. Many minds of a similar character are thus set upon the same train of thought and investigation. It is therefore natural that these similar inquiries should in at least some cases lead to the same conclusion. Truth is single; and if circumstances have placed earnest inquirers of the truth upon the narrow path that leads to it, they can hardly fail, although they may be unconscious of the presence of any other traveller, equally to reach their common destination.

We may also see why slavery is never inventive, and seldom even avails itself of the inventions of others. Not one of the conditions essential to the growth of inventions exists under this system. Slaves have neither the power nor the will to invent. They have not, and they are not permitted to acquire, at least in the modern form of slavery, the knowledge of natural forces which would enable them to make or to adopt inventions. As mere labourers their awkwardness and their carelessness are described as almost incredible. They can never be trusted about machinery; and are only fit under the strictest supervision for the coarsest kind of work.* Consequently, in a slave community where the bulk of the labour is entirely unskilled, an ingenious slave, even if no other obstacle intervened, could obtain no assistance in branches of art different from that in which he was engaged. A slave too has no property, and therefore no capital to assist his efforts. His master, so far

---

* See Olmsted's *Journeys in the Cotton Kingdom*, vol. i. p. 100.

from assisting or even encouraging his experiments, would probably regard them as excuses for idleness and as deserving rather of punishment than of reward. Above all, the slave feels no want. Even if under every disadvantage his constructive powers were successfully exerted, he knows that all the gain of his invention would belong to his master; and that any reduction in his own labour would merely afford an opportunity for imposing on him fresh toils. Among men thus unintelligent, uninstructed, unassisted, and uninterested, there is no room for invention.

The history of premature inventions is another curious subject which in like manner receives its explanation. It is true not merely that an invention is generally made when it is wanted; but when it is not wanted, even though it be made, it remains barren. Sometimes it is even wholly forgotten, and in a more auspicious moment is re-discovered and used. The conditions upon which its successful application depends have not been realized. Either from the difficulties of obtaining suitable co-operation, or from the expense or from the defective state of the arts involved in its working, there were no means, or no convenient means of carrying the invention into effect. The inventor has discovered a force too powerful for him or his contemporaries to utilize, or perhaps to appreciate. On the other hand, the pressure of the want may be slight; and the expense of carrying out the invention may be greater than in the then existing state of the demand it would be profitable to incur. In all these cases the knowledge has come too soon for any practical purpose. Thus Dudley was acquainted with the use of coal for smelting purposes; but his invention was in advance of the art of smelting as practised in his day. The Marquis of Worcester knew of the steam-engine, the Newcastle colliers had their railways, long before time and the hour raised up Watt. The Eastern nations were from remote ages familiar with the compass and with gunpowder; and yet these great agencies never produced in those nations the results which they have accomplished in the more congenial civilization of the West. Such too was the case with the great geographical discoveries of the world. Centuries before the achievement of the illustrious Genoese, the bold Scandinavian pirates had probably visited the shores of America. The circumnavigation of Africa was certainly known to the Ancients. These discoveries however remained dormant and forgotten until both of them were again made in an age better prepared to receive them. The expeditions of Columbus and of De Gama were due partly to the desire for exploring the surface of the globe after its form had been ascertained, and partly to the want then felt of new fields for commercial enterprise.* These undertakings were rendered practicable at that particular time by the increased use of the compass, and by the accumulation that had gradually taken place of the necessary wealth. While these conditions, the sense of the want, and the facilities for satisfying it, were absent, the knowledge of the fact remained idle; and ultimately sunk into complete oblivion. When they were fulfilled, the long-forgotten knowledge was speedily regained.

§ 8. Since invention implies the execution of human work by other means than human muscles, the circumstances of its beneficial application might at first sight seem indefinite. Its introduction would appear to be in every case pure gain. The fact however is far otherwise. Invention depends upon conditions that are always more or less costly. This preliminary or incidental cost may exceed the return afforded by the power that it sets in motion. The invention thus ceases to be an aid to labour, and is consequently laid aside. The first question that the commercial world asks

---

* Comte's *Positive Philosophy* (Miss Martineau's Edition), vol. ii. p. 385.

respecting any new invention is whether it will pay. For these purposes the material subject for consideration is whether, in the existing circumstances, the assistance that the new force is likely to afford will be sufficiently in excess of the cost of bringing it into operation. There have been at all times chimerical and silly projects which have afforded no unsuitable subjects for the humour of the satirist. But there have also been many real inventions which have shown the legitimate application of some new force, although in the circumstances its employment was not profitable. Examples of these unsuccessful inventions, where the cause of failure was so to speak not absolute but relative, abound in the records of patented processes. Many forces are known to be available for human purposes that are never so employed. The electric light is the most brilliant of all known means of illumination, but it is too costly for ordinary use. A fire which combines many advantages may be produced by the proper application of gas to platinum; but the platinum fire has not yet taken the place of the coal fire. Electricity and heat are capable of producing motion, and that too in such a manner as to impel machinery; yet, with the doubtful exception of the Ericson-engine, neither of these forces has proved commercially available; far less have they superseded steam. Nothing could at first sight have seemed more improbable than the success of the steam locomotive over the atmospheric locomotive. The power of the air, which was absolutely gratuitous, was proved to be capable of impelling railway carriages as effectually as the power of steam, generated by coals which were procured at a great cost and were brought from a considerable distance. But the conditions under which the force of the atmosphere could be applied were so onerous that the invention ceased to present the character of an aid, and its use has consequently been discontinued.

CHAPTER XII.

OF CO-OPERATION.

§ 1. The appearance of co-operation marks a new feature in our inquiry. An isolated labourer may procure for himself the assistance, limited indeed but still appreciable, of capital and of invention. But co-operation and exchange imply by their very terms the presence of other persons. These persons may be, and sometimes as in the case of the backwoodsman actually are, exclusively the members of the labourer's family. The family therefore, the lowest unit in the social system, exhibits a true co-operation although in the simplest and most rudimentary form. In that form this principle seems at first sight to present nothing peculiar to man. There are some cases where in even a higher degree than this, it is found among some of the lower animals. But except in a very few instances these animals exhibit no organization. Their union does not present that permanent concurrence in a general operation pursued by distinct and mutually subordinated means which is essential to association.* They merely repeat the same effort, and thus by the force of numbers obtain the desired power. Men, however,

---

* Comte's *Positive Philosophy*, vol. ii. p. 183.

have advanced a further step. They not only combine their labour, but they divide and apportion their several occupations. They can obtain an accumulation, not only of similar operations, but also of different operations, separately perhaps useless, yet so pre-arranged as to form a perfect and harmonious whole. By this means each person becomes dependent upon every other; and the character of industry, as the principle of co-operation extends, becomes proportionately complex.

The nature of co-operation requires little comment. It is not like capital or invention, liable to confusion or mistake. Nothing can be simpler than the combination of labour. A regiment of soldiers advancing together to the charge, a gang of navvies working at the same excavation, a number of compositors setting up type for the same book, all these persons exhibit an action that is perfectly plain and intelligible. When employments are divided, the process requires a little, and but a little, more attention. I do not now refer to that more complex form of co-operation which as society advances is gradually developed between different independent classes of the community. At present I am concerned only with that distribution of work which men deliberately and consciously adopt for the completion of a single object. The following example shows both labour combined and operations separated; and at the same time affords an additional illustration of the influence which capital and invention, when combined with each other and with co-operation, can exert.

In his celebrated chapter on the division of labour, Adam Smith takes one of his happiest illustrations from the woollen coat of the common labourer. He shows that this garment, coarse and rough as it may appear, is the produce of the joint labour of a great multitude of workmen. He notices the merchants and the carriers, the ship-builders and the sailors, and their thousand subsidiary arts, the complicated machines of the ship the mill and the loom, all which this homely production pre-supposes; and he points out the variety of employments requisite to form only "that very simple machine, the shears with which the shepherd shears the wool." This great natural organization of industry will receive a separate discussion. The present subject relates to a simpler form of co-operation, the existence of which might be at least conceived as possible, even if society were limited to these co-operators alone. I will therefore take a single stage in the preparation of the woollen coat. I will assume that the wool has been produced, and conveyed to the manufacturer: and I will trace the processes which in one of the great factories at Leeds, such a fleece, from the time that in its primitive state it enters the establishment until it is finished in the form of cloth and ready for sale to the cloth merchant, actually undergoes.* A single fleece sometimes contains as many as ten different qualities of wool. The wool is therefore in the first place sorted. It is next scoured with a ley and hot water: it is then washed with clear cold water; and is dried, first in a revolving machine called an extractor, and then by the action of steam. When the cloth is wool-dyed the process of dying succeeds. The matted locks are next opened and freed from dust by revolving cylinders armed with teeth. This process is twice performed; and the wool is then sprinkled with olive oil to to facilitate its working. Another machine removes any seeds of plants or grasses that adhere to the fleece; and after being scribbled, plucked, and carded, in three different machines, the wool makes its appearance perfectly opened, spread in a regular thickness and weight, and in cardings . or slivers of a light flimsy substance about three feet in

---

* *Journal of Statistical Society*, vol. xxii. p. 4.

length. These cardings are then joined at a frame called the billy, so as to make a continuous yarn; and are wound upon bobbins. The yarn is then spun on the mule; and is reeled, and warped, and put on the beam for the loom. The warp is sized with animal gelatine, and is wove. The cloth thus produced is scoured, to remove the oil and the size; and, when piece-dyed, is dyed. Any irregular threads, hairs, or dirt, are then picked out; and the cloth is milled, or fulled, with soap and hot water. It is then scoured, dried, and stretched on tenters. Its nap is raised by brushing it strongly on the gig with teazels fixed on a cylinder; and is cut both lengthwise and across with two cutting machines. It is then boiled so as to give it a permanent face; and afterwards brushed, and pressed in an hydraulic press. At this stage of the process the nap is cut a second time defects are removed; the manufacturer's name is stamped on the piece; and a second pressing takes place. It is then steamed, to take away its liability to spot; and finally it is folded for the warehouse. Every fleece thus undergoes thirty-four separate processes, for most of which separate machines and separate classes of workmen are required. The number of persons employed depends of course upon the extent of the business; the average is in a factory about twenty-six persons to each billy; and in a finishing-mill about eight persons to each gig.* Partly from this organization and partly from the use of better kinds of wool, the cloth of the present day is very superior to that made a century ago. The working man of the present day wears a better cloth than the fine gentleman could procure in the time of Adam Smith.

This co-operation extends from the simplest and commonest article to the most elaborate and fashionable. The labourer's coat is rivalled in the complexity of its manufac-

---

* *Journal of Statistical Society*, vol. **xxii**. p. 26.

ture by the lady's collar. "There are few manufactures which exhibit such division of employment as that of sewed-muslins. We have the spinning of the yarn for making the cloth; its warping and weaving and the reeling of the cotton for embroidery. We have next the designing and the drawing of the patterns, either on the stone or zinc plates, the block stereotype or copper-plate engraving, the printing of the patterns on the cloth, the despatch of the different pieces of printed cloth to at least 400 or 500 agents in Ireland ; the distribution of these throughout the country for embroidery, the return of these to the agents, and their transit back to Glasgow, their examination and preparation for the bleacher, the various operations through which they pass at the bleachfield, their return to the Glasgow warehouse, there to be made up, ironed, folded, ticketed, arranged according to quality and price, placed in fancy paper boxes, and packed ready to be dispatched either to the home or foreign market. The history of an embroidered collar or handkerchief could indeed tell as varied a tale as the famous adventures of a guinea." *

§. 2. Without some degree of the division of employment, as Mr. Mill remarks, the first rudiments of industrial civilization cannot exist. In proportion as that civilization advances, the practice of co-operation in both its forms steadily increases. The advantage is too great and the practice too general to have remained unobserved ; and there are accordingly few portions of Political Economy which have received such complete and ample consideration. Various illustrations have been offered to show the extent of the benefit which industry receives from this great auxiliary. None of them are or probably can be completely satisfactory ;

---

* *Journal of Statistical Society*, vol. xx. p. 426.

but, like all other such illustrations, they are useful in enabling us to form a more vivid if not a more exact conception of the subject than we otherwise could do. The two illustrations which on this subject may be almost considered classical, are the descriptions given by Adam Smith of the manufacture of pins, and by M. Say of that of playing cards. In the former passage, the best known perhaps in the *Wealth of Nations*, it is said that ten workmen can in concert make in one day 48,000 pins. The contribution therefore of each separate workman to this joint result is 4,800. But a man working by himself could not, we are told by the same authority, make twenty pins, perhaps not one pin, in a day. Assuming the highest amount thus suggested to be the fair day's work of the solitary artizan, the increase of industrial power thus effected by co-operation is as 240 to one. This great increase is obtained not by the substitution of any new force, or by the addition of any mechanical advantage; but merely by establishing a concert between the action of several men. In the example given by M. Say, the amount of the advantage is somewhat greater. That writer states that thirty workmen will produce in a day upwards of 15,000 cards, that is 500 cards for each man. If they were to work separately, each workman would probably produce at the utmost about two cards. Thus the produce of one day's labour of the thirty men would be, when they worked independently of each other, 60 cards, as against 15,000 obtained from their concerted exertions. In this case therefore the increase of power which mere co-operation affords is in the proportion of 258 to one.

Other examples less accurately known, but not less remarkable, will readily present themselves to every mind. One of these is the case of printing. A copy of *The Times* is said to contain about one million types. It is a good day's work for a compositor to set up 15,000. It would therefore take a single compositor upwards of eleven weeks of six days to set up, without corrections and without reference to its subsequent distribution, the type of a single number of that journal. The mere writing out of the copy for the press would also require a considerable time. That the desired result can be at all obtained is due of course to the combination of labour. But a printing office also presents ample illustrations of the division of employment. Besides the compositors, it finds occupation for superintendents of the whole office and of each branch, for the readers and their assistants, for the various persons who attend the machine and for messengers and other subordinates. There is also the broad distinction between those who print, and those who supply the copy to be printed. The earlier printers were generally scholars, and pursued the art for the purpose of extending the knowledge of their own works, or of the great authors whose writings they edited. It is almost impossible at the present day to realize the obstacles which authorship must thus have encountered. Even if we assume the existence of the type foundry, the paper manufacture, the printing-press and the steam-engine, each of them involving no small amount of co-operation, it would be a serious task for an author to print and pull his own copy, to fold stitch and bind the pages, and ultimately to effect the sale of the book. In the case of a newspaper, in addition to the powers of arrangement and the skill in composition that are required, an accurate knowledge of the most heterogeneous facts in the most remote places is essential to a first-class journal. Even if we leave out of consideration the co-operation involved in the services of those who are engaged in the transport of the intelligence, the collection of news from various parts of the same country, not to speak of different and distant countries, can only be effected by the combined action of many persons designedly working together, each in his separate department, for a common object. It is not too much to assert that the production of a single copy of *The Times* by a single labourer would occupy the whole duration of a long and active life.

Another example may be found in the perfection to which the practice of the artillery service has been brought. Those immense masses of metal which even with all the appliances that modern skill has invented for their use would in the hands of one man be quite unmanageable, are moved from place to place, levelled and discharged with a rapidity and precision attainable only by the most complete organization and the most practised skill. Each gun has its sergeant and eight attendants. Of these men the first points and commands : the second sponges : the third loads : the fourth points : the fifth fires : the sixth serves the third with ammunition : and the remaining three transmit from one to the other the ammunition from the distant cart until it comes to the hands of the sixth. A similar distribution of duties exists as to the limbering and unlimbering of the gun. A single gun thus served will fire two or three shots per minute, and can be removed to any required position with corresponding rapidity. Nine guns with nine separate men, if they could be fired at all, would not give more than one shot every three minutes : and if they were not wholly stationary would at least move with great difficulty and very slowly.

§ 3. The operation of a power so great and so characteristic of man deserves a careful analysis. The first and most obvious effect of co-operation is to increase the power of labour. This result, which it is needless further to illustrate, arises from the combination of labour ; and has its origin not in our reason only but in our sympathies. But when two men help each other, their reason quickly leads to the division of their occupations. This arrangement, to which the most important benefits of co-operation are due, holds out the prospect of two direct advantages. It saves labour, and it gains time. If two miners sink two adjacent shafts where one only would suffice, the labour of one of them is wasted. Had the second of them remained idle, the same result might have been attained. If this second man were to combine his labour with that of the other to sink the common shaft, the work would be done in at least half the time and with half the individual effort. If instead of directly assisting in the work, one of them were to supply some want of his companion, he would enable that companion to apply his whole time and powers uninterruptedly to the work. The principle on which this distribution of work depends is that it is frequently only the same or very nearly the same trouble to extend to several persons the service that is rendered to one. If a mining-shaft be sunk, a number of persons may use it with as much convenience as one. If gas or water-mains be laid down, the gas and the water can be supplied to a whole town almost as easily as they could to a single house. If a coach run or a ship ply between two places, the cost of conveying the full complement of passengers is the same as the cost for conveying only two or three. The fire and the water and the care requisite to prepare the food of one man will equally prepare the food of three or four. Consequently, where two men have to do two different things, if, in place of each performing these two several acts, they can with the same or nearly the same effort perform for their joint benefit each one act sufficient for the two, there is a clear saving of half their labour.

§ 4. The great primary advantage then of the division of employment is its economy of effort. With half the labour and in half the time that two men working separately require, the same two men when working in concert will produce an equal result. But this arrangement besides its great direct gain is found also to bring with it not a few unforeseen advantages. Men soon find that the repetition of the same act increases the facility with which that act is performed. A new motive therefore arises for continuing and extending an arrangement which had been commenced for other reasons. This motive is the increase, often to an extraordinary degree, of skill which the continued attention to the same occupation gives to the workman. This result of co-operation is probably the one which is best known, and most thoroughly appreciated. When a society has made any moderate advance, the division of employment is so far established that not merely is one man unable to do well the duty of another, but he is unable to do it at all. Even when a person has received a general training in different occupations, experience shows that he cannot compete successfully with the professors of the several specialities in which he has been instructed. In the older countries at least, the jack-of-all-trades, if such a person can still be there found, is now as much an object of contempt as his prototype* Margites was in days of old. We are so accustomed to the exhibition of that skill and dexterity both in mental and physical operations which practice gives, that we no longer regard it as marvellous; and accept its attainment as a thing of course. Yet if we reflect upon the skill which some men acquire and the consequent increase of productive power, we may well wonder at the extent of the debt which industry owes to co-operation. Some instances of increased skill have been already noticed, and innumerable other instances will readily present themselves to every mind; one however may be mentioned where this advantage admitted

---

* πόλλ' ἠπίστατο ἔργα κακῶς δ' ἠπίστατο πάντα.—*Margites.*

of a direct pecuniary measurement. It is said that a certain kind of gimp was reduced in price from three shillings to one penny, solely from the increased dexterity of the workmen.*

A serious inconvenience has been frequently noticed by economists as incident to the division of employments. The minute division of occupations is said to dull the powers and faculties of the worker; and while it gives him in one respect an extraordinary skill, renders him rather a living part of a complex machine, than a thinking man. "To have never done anything," observes M. Say, "but make the eighteenth part of a pin is a sorry account of a human being to give of his existence." It may however be questioned whether the account would be more satisfactory if its subject matter were not the fraction of a pin but an entire pin. Agricultural employments are less subdivided than those of trade or manufacture; and yet, an agricultural labourer is seldom as quick and as intelligent as a mechanic or an operative. It is rather the circumstances of his position than the character of his work that affects a man's powers. If the labourer's whole time be absorbed in one pursuit, whether it be a substantive or a subsidiary occupation, and whether it be intellectual or manual, whether it consist in making pins' heads or growing corn, or solving equations or writing Iambic verses, the excessive development which the faculties thus continually exercised will receive, is purchased by the inanition of the other powers. But the cause of this evil is not co-operation, but excessive labour. The remedy must be sought not in a reduction of the skill, but in a better distribution of the time and powers. We are not to destroy one set of faculties because its action has been excessive; but we should endeavour to bring into

---

* *Edinburgh Review*, vol. lxxxix. p. 81.

operation every part of our nature, and to establish the harmony of all.

§ 5. But co-operation effects another and a very important saving. The division of employment not only gives skill, but applies that skill to the best possible use. When a man is peculiarly skilled in any work, it is mere waste to employ him in a different kind. This remark is true even although his superiority in the other kind of work were manifest. If a man can in one day make a thousand pins heads or fifty nails, and if another man can make one hundred pins' heads or twenty-five nails, the former will if he be prudent confine himself to making pins' heads: and even though he be a better nailer than the other will abandon to that other the making of nails. A razor or a surgical knife will cut cloth; but the same office can be also performed by an humbler instrument. Prudence therefore suggests the appropriation of the costly steel for the purposes which it alone can fulfil. So too we may reasonably presume, notwithstanding Pope's sneer to the contrary, that Sir Isaac Newton could have added up some columns of figures when he was superintending the operations of the Mint; but that a prudent economy of his powers assigned this simple work to a clerk, and set the superior mind free for the performance of higher and more difficult duties.* This principle applies not only to different employments but also to minute subdivisions of the same occupation. Mr. Babbage† states that the operation of pin-making requires in its different parts such different degrees of skill that the wages earned by the persons employed vary from fourpence halfpenny a-day to six shillings. Consequently if the workman who is paid at

---

* See Macaulay's *History of England*, vol. iv. p. 704.
† *Economy of Manufactures*, p. 148.

that highest rate had to perform the whole process, the daily waste of his labour during the greater part of his time would be equivalent to the difference between six shillings and fourpence halfpenny. Without reference to the loss sustained in the quantity of work done, and supposing even that he could make a pound of pins in the same time in which ten workmen combining their labour can make ten pounds, the pins would cost in making three times and three-quarters as much as by means of proper organisation they now do. Mr. Babbage adds that in needle-making the difference would be still greater, for in that manufacture the scale of remuneration for different parts of the process varies from sixpence to twenty shillings a-day.

§ 6. One of the most marked of the secondary advantages resulting from the division of employment is its economy of time. This economy shows itself in various ways. Each man when the division of employment is established can apply himself uninterruptedly to his work. He has no occasion to leave his own occupation for the purpose of transacting other business elsewhere. Thus the time is saved which would have otherwise been lost in passing from one occupation to another, a consideration of obvious importance where the scenes of the two businesses are separated by any considerable interval. We must also take into account the loss, not very great perhaps in each individual instance but when accumulated distinctly appreciable, of the time spent in changing from one employment to another. Most persons when they are deeply engaged in any subject are unable at once to divert their thoughts sufficiently to take up with equal energy a different one. " A man generally saunters a little," as Adam Smith observes, " in turning his hand from one employment to another. This constant shifting of work generates a habit of careless application, and thus the workmen as we frequently see in remote districts becomes incapable of vigorous exertion even on the most pressing occasion." But even in cases where the operation occupies a small portion only of time, we may estimate the effect upon the gross result of any delay occasioned by a change of duties. It is said that one man working ten and another twelve hours will have made heads and points for 144,000 pins. But if one man were to make even with undiminished skill the head and the point of a pin alternately, and if we allow the loss of one second only each time he changed his work, the waste will amount in the 144,000 pins to eighty hours, that is to nearly four times the number of hours effectively employed.*

Another advantage in point of time is the shorter period of instruction which the division of employment requires. When each person is confined to a simple occupation or portion of an occupation, the time required for learning the art is much less than that which the acquisition of a variety of arts would involve. The workman therefore has both a longer period of his life in which he can exercise his skill and needs for his preliminary support and for his first experimental essays a smaller amount of food and clothing and a smaller amount of tools injured and materials spoiled than in a different state of circumstances he could expect.

§ 7. The presence of a second party which co-operation involves raises a question which in the case of the two preceding industrial aids had no place. This question relates to the terms upon which the parties consent to co-operate.

Men co-operate either voluntarily, or under compulsion. Where their action is voluntary, their inducement may be either mere sympathy and a desire of doing to their neigh-

---

* Longfield's *Lectures on Political Economy*, p. 99.

bors as they wish and expect their neighbors in similar circumstances to do to them; or an agreement either to share according to some stipulated proportion in the product of their joint exertions, or to receive in lieu of their remote and contingent claim an immediate and assured reward. With compulsory co-operation and with sympathetic co-operation I am not now concerned. The former is an abnormal phenomenon; the latter belongs to a different class of subjects from those which in these pages are discussed. Remunerated co-operation alone, whether the remuneration consist of an actual partition of the product or of a fixed money payment, is a subject of economic inquiry. Since the idea of force is thus excluded, and since co-operation implies a plurality of persons, the amount of this remuneration and all other circumstances connected with it must be determined by the joint will of the parties concerned. Some future object is sought: some combined action for its attainment is expedient: all the parties are to contribute to the attainment of that object: all are to share in the result. The object, the choice of partners, the contribution, and the ultimate division, are all matters of which the parties themselves are the exclusive judges.

But in the simpler forms of co-operation there is little room for discussion of terms. In the family, and even in the clan, co-operation is much more a matter of sympathy than of remuneration. When the latter principle is developed, the subject is complicated partly by the action of exchange, and partly by that peculiar modification of co-operation which takes place in an advanced state of society. It will therefore be convenient to reserve the further consideration of this subject until the principles of exchange and the modifying influence of society have been ascertained.

CHAPTER XIII.

ON THE CIRCUMSTANCES WHICH DETERMINE THE EXTENT OF CO-OPERATION.

§ 1. I have thus considered the forms which co-operation assumes, the extent to which it assists industry, and the mode of its action. The next subject of inquiry relates to the circumstances that are favourable or unfavourable to its development. No system of co-operation can be permanently successful unless it be sufficient, constant, and reliable. The number of persons expected to co-operate must be adequate to accomplish the proposed work. These persons must be able and willing to give their assistance; and must be available at the time and at the place at which their services are required. They must further be prepared to continue their exertions until the proposed work is completed. Assuming that the requisite number of persons is assembled where and when and for as long a time as it is desired, these persons must be able to rely upon each other for the efficient performance of their respective portions of the work. Again, as co-operation implies a number of workers, and as its results depend upon the harmony of their efforts, and as this harmony is attainable only by obedience to directions which can issue only from a single mind or at most from the joint agreement of a few minds, co-operation must be limited by the aptitude for discipline of those who are engaged in it. Further, co-operation depends not only upon the abilities of the several partners or upon their submission to control, but upon their fidelity to their engagements. The conditions therefore of extensive co-operation are a sufficient and constant supply of suitable labour; a mutual confidence in the several labourers that their exertions will be met by corresponding exertions on the part of all their comrades in the projected undertaking; and a readiness among all the cooperators faithfully to adhere to their engagements.

§ 2. The dependence of co-operation upon the number of persons co-operating is involved in the very term. Co-operation implies the concerted exertions of two or more persons. The extent therefore to which one person who desires to obtain the assistance of others can obtain that assistance must be limited by the number of those who are able and willing to assist him. The co-operation which he can obtain may be more or less extended in proportion to the greater or less number of those with whom he can act in concert. Accordingly in newly settled countries, where the population is both scanty and sparse, the want of labour is a prominent and continual grievance. The meaning of this complaint is that in those communities the opportunities for co-operation are deficient. There are many works which, as we have already seen, absolutely require for their execution the simultaneous exertions of a considerable number of persons. In the ordinary operations of seed-time or of harvest, in navigation and other means of transport, in many manufactures, in a thousand other operations, the want of such a number of labourers as the case requires is fatal to the success of the undertaking. Not fewer than 126 classes of occupations, each of them containing many varieties are enumerated as actually giving simultaneous employment to large numbers of workmen;* while in thirty-nine others the numbers engaged, though less, are still very considerable. Unless there be a sufficient number of persons to admit of their being advantageously carried on, these works cannot be at all attempted: and consequently no co-operation for the objects at which they aim can take place. Again, it is only in a large population that any extensive subdivision of employment is attainable. Without any present reference to the limit that the extent of the population as affording a market imposes, there must be a number of hands engaged when each occupation is divided into several distinct branches. The potteries for example contain from thirty to forty separate trades. The manufacture of woollen cloth requires as we have seen a similar number. Watch-making involves upwards of a hundred equally distinct branches. In the census returns of the cotton manufactures of Lancashire as many as 1225 separate occupations have been enumerated. It is obvious that the presence of the number of persons that correspond to the number of occupations that the trade in each of these cases requires, is a condition precedent to the separation of the respective occupations.

§ 3. The great increase of productive power which is thus obtained is therefore attainable in those circumstances only where there is a sufficient number of persons to undertake these various subdivisions of each employment. But there is another condition of co-operation which is of no less importance than the number of the labourers. Whatever the number may be, it is practically of little use unless the required number be forthcoming at the right time and in

---

* Levi's *Annals of Legislation*, vol. ii. p. 118.

the right place. " By far the greater part of the operations of industry," says Mr. Wakefield,* " and especially those of which the produce is great in proportion to the capital and labour employed, require a considerable time for their completion. As to most of them, it is not worth while to make a commencement without the certainty of being able to carry them on for several years. A large portion of the capital employed in them is fixed, inconvertible, durable. If anything happens to stop the operation, all this capital is lost. If the harvest cannot be gathered, the whole outlay in making it grow has been thrown away. Like examples without end might be cited. They show that constancy is a no less important principle than combination of labour. The importance of the principle of constancy is not seen in England, because rarely indeed does it happen that the labour which carries on a business is stopped against the will of the capitalist; and it perhaps never happens that a capitalist is deterred from entering on an undertaking by the fear that in the middle of it he may be left without labour. But in the colonies on the contrary, I will not say that this occurs every day because capitalists are so much afraid of it that they avoid its occurrence as much as they can by avoiding as much as possible operations which require much time for their completion ; but it occurs more or less to all who heedlessly engage in such operations, especially to new comers : and the general fear of it, the known difficulty of providing with certainty that operations shall not be interrupted by the inconstancy of labour, is as serious a colonial impediment to the productiveness of industry as the difficulty of combining labour in masses for only a short time."

§ 4. But if we suppose that there is a sufficiency of

---

* *Art of Colonization*, p. 169.

persons ready to co-operate at all required times, it still is necessary that during their period of co-operation the conduct of these persons should be reliable. Not only should the person who accepts the aid of others be satisfied that during their connection each man will continue to work as energetically as the rest. This confidence is far from being so common or its attainment so easy as we might at first suppose. The capacity of co-operation, as Mr. Mill* justly remarks, is the peculiar characteristic of civilized men. " In proportion as they put off the qualities of the savage, they become amenable to discipline ; capable of adhering to plans concerted beforehand and about which they may not have been consulted ; of subordinating their individual caprice to a preconceived determination, and performing severally the parts allotted to them in a combined undertaking. Works of all sorts, impracticable to the savage or the half-civilized, are daily accomplished by civilized nations, not by any greatness of faculties in the actual agents, but through the fact that each is able to rely with certainty on the others for the portion of the work which they respectively undertake." Nothing but frequent practice, and experience purchased in this as in every other instance by many losses and failures, will teach men the hard lesson of self-control. No other means is sufficient to develope in them those faculties and habits which taken together produce the capacity of co-operation.

Of the dependence of a skilled labourer on the exertions of his associates the following passage from Mr. Laing affords a good example.† " Whoever looks into the social economy of an English or Scotch manufacturing district in which the population has become thoroughly

---

* *Political Economy*, vol. ii. p. 261.
† *Notes of a Traveller*, p. 290.

imbued with the spirit of productiveness, will observe that it is not merely the expertness, despatch, and skill of the operative himself that are concerned in the prodigious amount of his production in a given time, but the labourer who wheels coals to his fire, the girl who makes ready his breakfast, the whole population in short from the pot-boy who brings his beer to the banker who keeps his employer's cash, are inspired with the same alert spirit, are in fact working to his hand with the same quickness and punctuality as he works himself. English workmen taken to the Continent always complain that they cannot get on with their work as at home, because of the slow unpunctual pipe-in-mouth working habits of those who have to work to their hands, and on whom their own activity and productiveness mainly depend."

§ 5. One great advantage of co-operation is the organization of which it admits. The direction of the work is left with the person who has the deepest interest in its success, or who is considered by those most interested in forming a correct judgment, to be the most capable of conducting it. This person is thus enabled to make the most convenient distribution of the work, and to concentrate on whatever point and at whatever time may be expedient the whole force of the workers. By this means, itself a form of the division of employment, the utmost efficiency, so far as regards the direction of the labour, is secured. But for the proper fulfilment of this arrangement several circumstances must concur. There must exist in those who are under direction proper habits of subordination. They must yield prompt and entire obedience to the orders they receive ; and after they have given to their superior the fullest information they possess, they must not suffer their own opinions of expediency to interfere with the zealous performance of apparently doubtful or inopportune tasks. The greater the number of co-operators, the more important does this subordination become. Examples of it occur in every large undertaking; but beyond all other instances its results are most conspicuous, and its practice is most rigidly enforced in the naval and the military services.

These habits of subordination are usually found in connection with other qualities affecting the general character of the worker. All those circumstances which determine the efficiency of labour relate in an almost equal degree to the capacity of co-operation. The circumstances most favourable for co-operation seem to be those in which the efficiency of labour is greatest, and in which habits of discipline at the same time prevail. It is only the intelligent worker that will appreciate the advantages of concerted labour. If co-operation be the great means of increasing skill, the skilled worker naturally seeks that co-operation by which alone his skill is made fully available. It is only among men whose efforts are habitually energetic that concert on anything like equal terms, and with its fullest results, can take place. It is only steady and upright men who will faithfully observe the terms of their engagement. The increase therefore of intelligence, of skill, of habits of energy and of good conduct, tends proportionately to increase, just as their deficiency tends to diminish, the aptitude for co-operation. It is one of the principal evils of slave labour that the master can never rely even in the most trifling particulars upon the execution of his orders. The stupidity, the carelessness, and the perversity recorded of slaves are almost incredible. If the master's back be turned, the most valuable animal will be injured or neglected; the most valuable commodity will be destroyed or spoiled. It is a common remark of slave-owners that "you can make slaves work, but you cannot make them think." The negro, it is said,* "can never be safely employed in factories: he can never be trained to exercise judgment: he always depends upon machinery doing its own work and cannot be made to watch it. He neglects it until something is broken or there is great waste." "In working niggers," said a planter to Mr. Olmsted, "we must always calculate that they will not labour at all except to avoid punishment, and they will never do more than just enough to save themselves from being punished, and no amount of punishment will prevent their working carelessly and indifferently. It always seems on the plantation as if they took pains to break all the tools and spoil all the cattle that they possibly can, even when they know that they will be directly punished for it."

§ 6. Besides the reliability and the docility of the parties, there are other kindred qualities of the highest importance to successful co-operation. The co-operators must be not only suited to each other, and obedient to directions; but must also be desirous honestly to fulfil their respective engagements. "It is well worthy of meditation," says Mr. Mill,† "how much of the aggregate effect of men's labour depends upon their trustworthiness. All the labour now expended in watching that they have fulfilled their engagements, or in verifying that they have fulfilled it, is so much withdrawn from the real business of production to be devoted to a subsidiary function rendered needful not by the necessity of things, but by the dishonesty of men. Nor are the greatest outward precautions more than very imperfectly efficacious where, as is now almost invariably the case with hired labourers, the slightest relaxation of vigilance is an opportunity eagerly seized for eluding performance of

---

* Olmsted's *Journeys in the Cotton Kingdom*, vol. i. p. 100.
† *Political Economy*, vol. i. p. 135.

their contract. The advantage to mankind of being able to trust one another penetrates into every crevice and cranny of human life : the economical is perhaps the smallest part of it, yet even this is incalculable." After noticing some instances of wealth wasted by improbity, he continues, "As the standard of integrity in a community rises higher, all these expenses become less. But this positive saving would be by far outweighed by the immense increase in the produce of all kinds of labour and saving of time and expenditure which would be obtained if the labourers honestly performed what they undertake, and by the increased spirit, the feeling of power and confidence, with which works of all sorts would be planned and carried on by those who felt that all whose aid was required would do their part faithfully according to their contracts. Conjoint action is possible just in proportion as human beings can rely on each other."

In addition to the cases cited by Mr. Mill from Mr. Babbage in illustration of the advantage to society of integrity, and the waste occasioned by dishonesty, there is an obvious example in the high rate of interest charged upon loans made upon personal security only to persons whose reputation is doubtful. So too the poorer classes pay unusually high rents, partly indeed from the increased trouble of collecting a number of small sums, but principally from the difficulty and uncertainty of obtaining payment. The effect of this distrust is most clearly seen where the law provides no remedy or no sufficient remedy for breach of contract. By the usury laws all contracts in excess of the legal rate of interest were rendered void. The rate of interest on all such illegal transactions consequently rose far beyond what the ordinary market price would have been, not only because the number of lenders was thus limited, but also to cover the losses consequent upon the bad debts of which payment could not be enforced. On the other side also there are many and more pleasing illustrations. In parts of Switzerland* watches and portions of watches are readily entrusted by the manufacturers to workmen of whom they know but little, and no loss ever ensues. Among the London costermongers, it is estimated † that property to the amount of £10,000 is exposed every night without any danger of loss as far as their own members are concerned. But for an instance of confidence as affecting almost every transaction of society it is not necessary to go beyond the simple case that Mr. Burton‡ puts " of a person receiving payment in Scottish one-pound notes who after having counted the pieces of paper in his hand would probably be quite unable to tell whether they were all the notes of one bank or each from a different bank. Having counted them, doubled them up, and nodded his head, he has absolved his original debtor, and taken obligations from the bankers one or more who have issued the notes put in a bundle in his pocket-book ; and yet he cannot tell the name of any one of the individuals who are thus made his debtors, he does not know their position in society, he does not even know the designations of the establishments from which the notes have issued."

§ 7. Such are the circumstances favourable to the development of co-operation. The next inquiry relates to the extent to which that development may proceed. Co-operation is an aid to labour. Its practice therefore will continue so long and so long only as it retains its character

---

* *Reports of Secretaries of Embassies and Legations*, 1858, p. 150.
† Mayhew's *London Labour*, vol. i. p. 26.
‡ *Political Economy*, p. 269.

of an aid. It becomes excessive when in the case of combined labour there is a disproportion between the number of workers and the quantity of work. Either more men are employed than are needed for the execution of the work, or the quantity of work executed is greater than the wants of the persons for whose benefit it is intended require. The former is the familiar case of too many servants being employed in a house, too many workmen in the construction of any work, too many clerks in a public office. The latter case occurs where in consequence of a glutted market, factories are put upon short time. The test of an excess in the number of co-operators in any occupation is very simple. There are enough and not more than enough workers when each person, in a given condition of production, is fully occupied in his own business. If the time of any of the workers be not wholly occupied, the co-operation is to that extent excessive. Of such an excess the combination of labour presents the simplest instance. When the work of an office can be performed by four clerks, it would be not only useless but burdensome to employ ten. But there may be an excessive division as well as an excessive combination. "If by the separation of pin-making into ten distinct employments 48,000 pins can be made in a day, this separation will only be advisable if the number of accessible consumers is such as to require every day something like 48,000 pins. If there is only a demand for 24,000 the division of labour can only be advantageously carried to the extent which will every day produce that smaller number."* This excessive distribution is generally the consequence of the excessive combination. Where too many persons act together, they almost instinctively try to find some duty for each of them, and so give some colour of use to their joint action. The

---

* Mill's *Political Economy*, vol. i. p. 160.

system of castes in India, where almost every domestic service has its appropriate minister, is a well-known example of excessive combination, and consequently excessive division. In that state of society where numerous retainers are maintained by a single chief, the same principle is apparent. In the earlier part of the present century it was the custom of the Russian nobles so to maintain great multitudes of servants.* In some of the larger palaces of Moscow upwards of a thousand domestics were assembled. No nobleman of any importance was content with a smaller retinue than twenty or thirty. It was the great problem of Russian domestic life to find some pretence of employment for these people. Thus one servant had nothing to do but to sweep a flight of stairs: another carried water for use at dinner: the water that was drunk during the evening was the charge of a third, and the whole business of the household was in this way portioned out with ludicrous minuteness. There are also cases in which although the co-operation cannot be said to be excessive, its burden presses close upon its advantage. Such a result takes place when the function is unavoidable, but is of rare occurrence, or relates to few objects. It is for example necessary to employ an engineer in every factory, but when there is only a single factory, unless the number of machines be so great as to require his undivided attention, such a necessity is a serious burden. It is for this reason among others that it has been found in practice so difficult to establish factories in localities remote from the recognized seats of that kind of industry. In like manner, it is often absolutely necessary to station a police magistrate or other public officer in some remote locality, although his time may not be there half occupied. Hence the cost of governing a

---

* Haxthausen's *Russian Empire*, vol. i. p. 48.

new country when estimated by the number of its population is generally out of all proportion to the cost of analogous services in an older community. For such cases of excessive co-operation there are two remedies. Either the number of workers may be reduced, or their work may be increased. The one course checks and limits the division of employment; the other tends to extend it. This latter course is only available where the wants of the persons beneficially interested in the result admit of the increased supply which it involves; and this expansion of wants in relation to a single object can only take place where the surplus of that object, after the owner's desires are satisfied, is convertible into objects of a different kind. This conversion is effected by exchange. Thus as all the leading economists have taught, the extent to which co-operation may be carried depends upon the power of exchanging, or in other words upon the extent of the market.

§ 8. It is a question of some importance to ascertain the classes of cases in which the excess of co-operation is most frequently observed. The answer to this question has been to some extent anticipated. The most common and the most important case is that in which the demand for the product is small. If by their united exertions two or more men produce a greater quantity of any particular object than they require for their own use, and have no means of converting the surplus into something that they do want, or if they require some particular object which they cannot obtain by purchase or otherwise from others, they will in the one case cease to expend unprofitable labour, and in the other will themselves supply their wants as best they may. Accordingly when the population is poor, or sparse, or scanty, or destitute of facilities for exchange with other communities, in other words when the market whether external or domestic is deficient, the division of employments will be confined within very narrow limits. In such circumstances men will supply their various wants, imperfectly indeed but still sufficiently, by their direct labour. If a man so situated were to confine himself to one occupation, the proceeds of his labour would be greater, but would neither themselves satisfy nor procure the means of satisfying his requirements.

There are also some occupations which, as the necessity for performing them is not constant but is of periodical recurrence, do not admit of any extensive division. The principal example of this kind is found in agricultural pursuits, although even in these the recent introduction of machinery seems to be gradually tending to the formation of distinct trades. In those countries where the climate confines people for a considerable part of the year within-doors, there must also be some change of occupation. It is from this desire of employing their leisure hours, whether that leisure be occasioned by physical or by economic circumstances, that those manufactures which are known as domestic have taken rise. On the other hand, when the condition of the market admits of the use of extensive and complicated machinery, each part of which requires attention and for which materials must be constantly prepared, the tendency towards the separation of occupations is strong. This tendency, as Mr. Mill* has remarked, is favoured by an increase in the quantity of capital, and by the existence of capital in a few hands. In the former case, new branches of industry are more easily formed by the investment of new than by the displacement of old capital. In the latter, one person more readily recognizes his advantage, and more steadily pursues it, than a number of persons of different degrees of intelligence, and sometimes with conflicting purposes. But

---

* *Political Economy*, vol. i. p. 174.

even among a number of small capitalists, if a distinct perception of the advantage be once obtained, there will be a perfect readiness to co-operate, and the frequent exercise of the act in one set of circumstances will naturally facilitate its adoption in a different set. Many interesting examples may be found in the pages of Mr. Mill and Mr. Carey, more especially the well-known case of the dairies on the Jura, and the various co-operative acts among the peasant proprietors in the New England States and in different European countries.

§ 9. Another limit to the use of co-operation is found in its magnitude. In proportion to the increase in the number of the co-operators, their motives for exertion grow weak. The greater the number of persons interested in any product, the less the share of each of them must be. That limited liability which to the capitalist is the great charm of a co-operative enterprise, implies also a sense of limited responsibility. If it diminish the penalties of failure, it thereby diminishes the security against that conduct which leads to failure. The whole of the reward does not attach exclusively to the deserving; the whole of the punishment does not fall upon the indolent. There are indeed methods by which these difficulties, although they cannot be wholly removed, may be at least reduced. The managers of a large association are usually more intelligent and more skilful than the average, or even the great majority, of their individual rivals. The organization of their work is from the very nature of the case better than a smaller business can command. But though ability may be hired, zeal never can. In all such associations the great natural stimulus to unremitting industry is wanting. No contrivance that has yet been invented can supply the place of the feeling that the workman is labouring not for another but for himself;

and that it is exclusively upon his own behaviour and his own exertions that his success or his ruin depends. Accordingly in those cases where the magnitude of the undertaking is not beyond the reach of individual enterprise, experience shows that companies are most successful in those occupations in which the influence of these self-regarding motives is least felt. There are many occupations which are reducible to a fixed routine, and consist in the exact adherence to known and definite rules. The observance of these rules forms an obvious moral obligation on the part of one who has undertaken to carry them into effect: and this obligation may further be enforced by loss of employment, and the discredit that attaches to neglect. It is also usual in many of the best managed companies to quicken the zeal of their officers by making a portion of their remuneration to consist in a per centage of the profits of the undertaking. "The personal interest," says Mr. Mill,* "thus given to hired servants is not comparable in intensity to that of the owner of capital, but is sufficient to be a very material stimulus to zeal and carefulness, and when added to the advantage of superior intelligence often raises the quality of the service much above that which the generality of masters are capable of rendering to themselves."

"The only trades," says Adam Smith,† "which it seems possible for a joint-stock company to carry on successfully without an exclusive privilege are those of which all the operations are capable of being reduced to what is called a routine, or to such an uniformity of method as admits of little or no variation of this kind, as first the banking trade; secondly, the trade of insurance from fire and from sea risk and capture in time of war; thirdly, the trade of making

---

* *Political Economy*, vol. i. p. 173.
† Book v. chap. i.

and maintaining a navigable cut or canal; and fourthly, the similar trade of bringing water for the supply of a great city." This celebrated passage Mr. Mill regards as the overstatement of a true principle. In his view Smith both failed to take sufficient notice of the countervailing considerations which operate in favour of associations, and formed too hasty a generalization from the limited experience of his own day. Yet there seems reason to believe that that great thinker justly appreciated the general principle that governs this case, although the advance of society has enlarged since his time the list of instances which that principle comprehends. Some years since a distinguished economist investigated this subject, and carefully analysed the existing statistics of joint stock companies.* Since that time however considerable additional experience has been obtained. Unfortunately these inquiries do not seem to have been continued, and I have not had the opportunity of myself pursuing them. But it appears that from the year 1845 to 1850 inclusive 2,885 companies were started, and were provisionally registered under the provisions of the Joint Stock Companies Act. Of these, about 300 railway companies obtained special Acts of Parliament which superseded the necessity of complete registration. But of the remaining 2,585 companies only 455, or about one in every five, were fully registered. It was a provision of this Act that every company carrying on business should make an annual return and receive an annual certificate. Neglect to comply with this provision was rendered penal: and it is therefore reasonable to infer that the number of annual certificates fairly represented the number of companies in actual operation. The total number of certificates so taken was in

---

* *Transactions of Dublin Statistical Society*, vol. iii. *Statistics of Joint Stock Companies*, by J. A. Lawson, LL.D., p. 7.

1850 only 304. These successful companies consisted principally of assurance companies, railway and canal companies, steam-navigation companies, gas and water works companies, and banking and mining companies. So closely did the results of modern experience coincide with the teachings of Adam Smith.

I am far from contending that there is any absolute impossibility in the success of a company of a different kind from those thus specified. The fair inference seems to be that in our present circumstances companies of this character have a reasonable prospect of success, while the chances are considerable against the success of companies that do not come within Adam Smith's principle. But in some occupations not so included, improved methods of business may introduce some portion of routine; and in others a greater capacity for co-operation, and greater skill in its practice may reduce the loss that arises from diminished zeal. Many attempts have been made, and in some cases successfully, to extend the range of associated enterprises. Of these attempts the most important have been made by manufacturing operatives who desired to establish factories for themselves. The movement has been checked in France, where it originated, by the interference of the Government. But in England numerous associations of this kind are now in operation. The earliest and most remarkable of them is the Rochdale Equitable Pioneers' Co-operation Society. This association which has flourished since 1844 is described by a writer* familiar with its history as having for its characteristic principles the subordination of the material well-being of its members to their social and intellectual advancement, and the restriction of all their dealings to ready-money transactions exclusively. But the main cause

---

* *Journal of Statistical Society*, vol. xxiv. p. 514.

of its success appears to be that the governing body is kept under the constant and rigid superintendence of a proprietary resident on the spot and practically acquainted with the nature of the operations carried on with their capital. The zeal of the co-operators is thus stimulated as far perhaps as in such circumstances it can be : the countervailing agencies are in full operation : and the novelty of the enterprise and its supposed importance in other respects besides that of a commercial speculation are doubtless not without their influence.

I cannot refuse my sympathy to such attempts or withhold my admiration from the energy and the skill with which they have sometimes achieved success. They have however suffered in public estimation from their supposed connection with socialistic principles. But whatever may have been the views of some earlier supporters of these associations, no such connection necessarily exists. On the contrary, where associations were founded on these principles, the operatives seem to have quickly discovered the practical inconvenience of such doctrines. "With wonderful rapidity," says Mr. Mill,* "the associated work-people have learnt to correct those of the ideas they set out with, which are in opposition to the teaching of reason and experience. Almost all the associations at first excluded piece-work, and gave equal wages whether the work done was more or less. Almost all have abandoned this system : and after allowing to every one a fixed minimum sufficient for subsistence they apportion all further remuneration according to the work done : most of them even dividing the profits at the end of the year in the same proportion as the earnings." Associations so conducted are indeed beset with practical difficulties, but they do not violate any economic principle. It is not against

---

* *Political Economy*, vol. ii. p. 350.

co-operation, but against communism that economic science protests: not against the increase that association gives to the efficiency of labour: but to the diminution of that efficiency by the interference with individual freedom of action and by the weakening of the motives which lead to the acquisition and accumulation of wealth. No reasonable objection can be taken to the theory of these associations. That theory consists in the extension to the working classes of the principle familiar to the trading classes of partnership and remuneration according to the service rendered. The successful application of this principle is to a great extent a question of time and of experience. Men who have a sufficient capacity for co-operation may successfully conduct a joint undertaking which would be beyond the powers of less skilful persons. Nor can there be a doubt that the mere attempt to carry on these associations affords, as Mr. Mill observes, "a course of education in the mental and moral faculties by which alone success can be either deserved or attained." While therefore on jural and political grounds, such associations ought not to meet any legal impediments, but on the contrary ought to receive from the law the utmost facilities for the transaction of their business and the enforcement of their rights, the economist should regard their success, not as a violation of an admitted principle, but as the sign of improved associative skill, and consequently as a positive gain to the action of a great industrial aid.

CHAPTER XIV.

OF EXCHANGE.

§ 1. I now approach the consideration of that great agent which with an excusable exaggeration some writers have regarded as the sole subject of economic science. Although a less exalted rank has been assigned in these pages to the theory of exchange, this lower view of its position does not proceed from any insensibility to its influence. Coming into at least full operation at a late period of industrial development, exchange quickens into new life as well the primary elements of production as its own fellow industrial auxiliaries. It suggests to the labourer new wants. It at the same time provides the easiest means of satisfying these wants. It enables the ignorant and the weak innocently to profit by the learning of the wise, and the vigour of the strong. It extends to the inhabitants of different regions a share in those natural advantages of which nature seems to have granted to each region the exclusive possession. It affords larger means for the accumulation of capital, and an ampler field for its profitable occupation. To its demands for increased facilities both of production and of intercourse some of our most important inventions are due. Above all, it is the complement and the crown of co-operation, carrying out the full effects of that great auxiliary to an extent that would otherwise be impracticable, and establishing not only between unacquainted individuals but remote and jealous nations an unpremeditated and almost unconscious, yet not the less complete or effectual, system of association. Nor must we omit, although they hardly come within the range of the present inquiry, the indirect benefits of exchange. It enlarges the sphere of men's observation and so of their knowledge. It substitutes for the ferocious antipathies of an earlier age the friendly relations which spring from a sense of reciprocal advantage. It renders both parties dependent on each other for a multitude of their daily enjoyments, and thus binds them in a sort of unacknowledged yet powerful frankpledge. Disputes consequently whether private or national tend to become more odious in conception and less facile in execution. Thus by a method that the most obtuse cannot overlook, and the most wilful cannot misunderstand, it teaches the great moral lesson that the benefit of each is the benefit of all; and that a wrong done to one class is sure to extend its influence over the whole community.

Exchange may be described as the voluntary transfer by one person to another of one instrument of enjoyment in consideration of the reciprocal transfer of a different one. There must therefore be two parties to every exchange; and there must also be a consideration. It is this latter circumstance which distinguishes exchange from other forms of giving and receiving. A free gift is something different from an exchange: so is a robbery: so too is a tribute. All these transactions fail in the essential condition of reciprocity. There is no *quid pro quo*. But as in an exchange there is something received, so also there is something given. There are two parties, each of whom is influenced by the desire of enjoyment and the dislike of effort; and to each of whom the transaction presents itself in a different light. Each has an inducement and a sacrifice; but the inducement and the sacrifice are reciprocally inverted. Each obtains at a smaller cost than he otherwise could the means of satisfying a desire or of accomplishing a purpose: and each therefore finds, or expects that he will find, the exchange beneficial. From these considerations the conditions of exchange may be deduced. In every such transaction there must be enjoyment; and in like manner there must be cost. There must also be a proportion between that enjoyment and that cost; and that proportion must be more or less in favour of the enjoyment. Further, as there are two parties to every exchange, the principles now stated must apply equally to each of the two, and therefore the transaction must be, at least in their opinion at that time, beneficial to each.

§ 2. Some attempt however imperfect may be made to illustrate by a numerical expression the extent to which exchange reduces cost. "We annually import into this country," says Mr. Senior,* in a work written nearly thirty years ago, "about thirty million pounds of tea. The whole expense of purchasing and importing this quantity does not exceed £2,250,000, or about 1s. 6d. a pound, a sum equal to the value of the labour of only 45,000 men, supposing their annual wages to amount to £50 a year. With our agricultural skill and our coal mines, and at the expense of above 40s. a pound instead of 1s. 6d., that is at the cost of the labour of about one million two hundred thousand men instead of forty-five thousand, we might produce our own tea and enjoy the pride of being independent of China. But one million two hundred thousand is about the number of all the men engaged in agricultural labour

---

* *Political Economy*, p. 76.

throughout England. A single trade, and that not an extensive one, supplies as much tea, and probably of a better sort, as could be obtained if it were possible to devote every farm and every garden to its domestic production."

A curious calculation has been made* to show the loss formerly entailed upon the agricultural classes in France by the prohibitory duty then imposed on English iron. The lands then cultivated in France amounted to about fifty-seven millions of acres. From the quantity of land that a team of oxen could plough, the number of ploughs required for the cultivation of such an extent of land was taken to be a million and a half. The annual use and waste of iron on each plough has been estimated to be from forty to fifty kilogrammes. The whole consumption therefore at ninety francs per kilogramme amounts to two millions seven hundred thousand pounds. Although this estimate is said to be too high for an average calculation, the iron could undoubtedly have been imported at half the price; and the annual loss to agriculture alone must be taken at above one million sterling. The indirect loss arising from the use of wooden ploughs or iron ploughs of an inferior description cannot perhaps be computed; but it probably represents a far greater sum.

§ 3. It is not difficult to understand the nature of the assistance which exchange affords, and the mode of its operation. Exchange is an interchange of superfluities. Men give objects which they desire less, in return for objects which they desire more. The parties to an exchange have their superfluities and their deficiencies inverted, because their natural powers are different; because their acquired powers are not less varied; because the natural agents

---

* Porter's *Progress of the Nation*, p. 289.

within their reach from differences of time place or circumstance are also different; and because the opportunities of these parties and their appliances for dealing with these natural agents present still greater degrees of diversity. Without the aid of exchange the satisfaction sought could in many cases either never be obtained or be obtained only at the cost of a greatly increased effort. No possible effort on his part would enable a hodman to perform on himself an important surgical operation; or to ascertain and establish his rights in a complicated suit in equity. No physical training on the other hand would enable a weak and puny clerk to carry the load of a porter or a coalheaver. No artificial contrivance could raise tin in Durham or coal in Cornwall. Spices or pineapples may be grown in the British Islands; but only under such conditions of cost as would place them beyond the reach of ordinary consumers. In tropical countries these articles grow in abundance; and the inhabitants part with them for so small a consideration that even after a long sea carriage they can be sold in England at a very moderate price.

Exchange not merely acts as an economical form of production, but without it a large part of our ordinary production would never have taken place. It implies the existence in the purchaser of wants previously ungratified or unknown, and the power of rendering his labour or his property available for their satisfaction. It thus utilizes, or at least increases the utility of, labour or of property that otherwise would have been useless, or at the most have afforded an incomplete enjoyment. When a savage discovers that the stones, or the skins, which are to him so common as to claim no special attention, can obtain for him beads or nails or other objects of his desire, he learns to prize and collect those commodities. When a labourer who wastes over his work the time for which he received no appreciable reward, finds full employment at piece-work, every minute acquires in his estimation a very different importance. When we hear that trade is brisk or markets lively, we know that capital is well remunerated. In all these cases the new value attached to the stones or the peltry, the increased reward for labour, the rapidity with which the capital is turned over, indicate a new demand. This demand may be caused either by an absolute increase of the population, or by an increase in the number of the buyers arising from increased facilities for exchange. This new demand gives to the vendors an increased power of satisfying their wants; and incites them to new or to increased industrial efforts. It supplies them with a motive, or more accurately it places them under industrial conditions, that did not previously exist. It is this effect that both many economists and the ordinary forms of speech indicate by such phrases as the finding a new market, an outlet for the surplus produce of the country, and similar expressions. They do not mean to deny that the benefit of an exchange to a purchaser consists in the incoming, not in the outgoing, in that which he acquires, not in that with which he parts. The expression is the same loose form used in the ordinary demand for employment, that is for labour not indeed for its own sake, but for the sake of the reward that accompanies it. These terms merely imply that the labour which was previously idle or comparatively inefficient, now becomes remunerative, or more largely remunerative than before. Exchange then has a double function. It both aids labour, and it stimulates it. It not only reduces to a minimum the effort required to satisfy any want; but it excites certain efforts which otherwise would never have been made.

§ 4. Since exchange is a method of procuring upon advantageous terms some means of satisfying a want, the object which a man seeks to acquire by exchange must be, or must appear to him to be, in some manner desirable. It is immaterial whether the object be really capable of gratifying the desire or of accomplishing the purpose; or whether the desire or the purpose be one which a good or a prudent man should entertain. It is sufficient if the object be one which the purchaser, whether rightly or wrongly, desires to possess. But as exchange professes only to reduce, not to abolish trouble, as it necessarily implies a consideration, it follows that the desirable object must also be one which is difficult of attainment. The cause from which this difficulty proceeds, is as little material to the exchange as is the nature of the desire. It is enough that the difficulty does in fact exist. It was at one time said by many economists that value depended on limitation of supply. But scarcity is only one out of several causes of difficulty. An object may be difficult of attainment, not only from its absolute rarity, but from the time or trouble that its appropriation requires, or from the general inability or reluctance from whatever cause to render the particular service. In a Canadian forest for example a man may obtain any quantity of timber by merely cutting it down, or any quantity of wild strawberries by merely picking them up; yet the trouble and the inconvenience of wood-cutting or of strawberry-picking are so great, that men are always willing to pay a price for having the work performed for them by others.

Again, as the peculiar method used in exchange of reducing trouble is transfer, it follows that those desirable objects which are difficult of attainment if they be subjects of exchange must also be transferable. To say that an exchangeable object must be transferable seems almost equivalent to the proposition that an exchangeable object must be exchangeable. Where the transferability is natural, the idea of exchange never arises. The heat and light of the sun, personal health and beauty, mental endowments, although both desirable and by any human exertion unattainable, are never in the market. We can purchase their use or their results; but the things themselves cannot be transferred. It is the artificial obstacles to transfer that direct our attention to this condition. Of these obstacles I shall elsewhere have occasion to treat. At present it is enough to indicate the great quantities of land which under family settlements are for many years together inalienable. There are also many unmarketable titles to real property. The holder of such property may be sufficiently secure from dispossession; but his title is open to dispute, and purchasers will not buy a lawsuit. He can thus comply with two conditions of exchange, but not with the third; and consequently his property is for the time not an exchangeable object. The value of the land thus rendered inalienable under the old system of real property law in England and Ireland was enormous. Some estimate of its extent may be formed from the fact that the Commissioners of Incumbered Estates in Ireland sold land which would otherwise have been inalienable to the amount of upwards of twenty-three millions of money.

There is yet another condition of exchangeability. The transfer of one such desirable object as that already described for another such object implies that the two objects are dissimilar. There is no motive for the interchange of objects that are precisely alike. A man parts with his property, or exerts his labour, to obtain something which he could not on equally advantageous terms otherwise obtain. If he receive in return something substantially the same as that which he gave, he has had all the trouble of the transaction without any of the expected advantage. At best the exchange, if it be not an actual burden, has ceased to be of any use. Accordingly we find that it is between men in different occupations, and not between men whose occupations are the same, that the great bulk of exchanges takes place. A carpenter seldom deals with a carpenter, or a mason with a mason; but they both deal with the butcher and the baker. No miner seeks to change his gold for other gold; and no farmer buys wheat with similar wheat. Diversity is essential to exchange. Like the condition of transferability, this condition is so involved in all our ideas of exchange that any illustration or enforcement of it seems absurd; and even its formal statement sounds like a truism. But although it is so obvious in its simple form, it brings with it consequences that are not equally apparent.

In every such transaction both parties must be free agents. Each man judges for himself of the comparative merits of the proposed inducement and the proposed cost. The idea of compulsion at once suggests a disturbing element. Every true exchange is voluntary. This limitation of exchange would not have required any special notice, were it not that an eminent writer* has denied its accuracy. It has been urged that if this view were correct the expressions "voluntary exchange" or "forced exchange" would be improper. But the impropriety consists rather in the undue extension than in the undue limitation of the term. It is only by a kind of metaphor that we can speak of a forced exchange. That expression denotes something that has the appearance of an exchange but yet is different from it. Presents made on the tacit understanding that an equivalent will be returned are not really gifts. So a transaction without such a consideration as the party interested is willing to accept may perhaps be a theft, but is not really an exchange. Taxation has been cited as an instance of involuntary exchange. But a tax is the equivalent for the service of protection rendered. In a

---

* Archbishop Whately's *Lectures on Political Economy*, p. 10.

free State, either an endowment in the form of an hereditary revenue is provided by the original constitution of the country, or the people by their representatives consent to and determine the amount of their payments. In less favoured countries, the Government it is true both determines the remuneration and judges of the quality of the service. I am not however concerned to justify the practice of such governments, or to defend them from any charge that their opponents may bring against them. It is no objection to the true meaning of the term either that it is sometimes coupled with superfluous or incongruous epithets: or that it like many other terms bears its silent but emphatic testimony against the injustice of some rulers of mankind.

§ 5. In every exchange there are two points to which a purchaser must look. He must consider the strength of his desire, and the cost at which he either by his own act, or by the agency of others can gratify that desire. As either party in an exchange is in turn vendor and purchaser according to the point of view from which we regard him, the same considerations influence each. The amount of any other exchangeable object which any object can command in exchange is said to be the value of that object. When this value is expressed in money, it is called price, a term which as it is free from the ambiguities inseparable from value I shall habitually use. Although in every exchange the conditions both of desirability and of difficulty must co-exist, yet that of difficulty is generally the more active. The strength of the desire can only be tested by the difficulty which it will overcome. Without such a test, its force remains unknown even to the person who feels it. If the difficulty exceed the purchaser's desire, no exchange can take place. The motive for the exchange, the saving of sacrifice, will have ceased to operate. The point then at which the sacrifice is felt to be fully equal to the enjoyment, is the extreme limit which price can reach. The price may fall short of this point by any degree, but can in no case exceed it. It is however in rare and exceptional cases that price even approaches this limit. For although the excess of difficulty over desirability is fatal to exchange, the opposite condition of an excess of desirability is most conducive to it. In such circumstances the motive to exchange is in full operation; and the only subject of doubt is the precise amount of the difficulty. This question is determined by the mutual consent after free discussion of both parties. The purchaser considers the convenience to him of the subject of the transaction, and the difficulty he would have in otherwise procuring it. The vendor considers the convenience to him of the price offered, and the difficulty he would have in otherwise procuring that price. Neither of the two has any concern with the convenience of the other. Each seeks merely the smallest possible amount of sacrifice for himself. But exchange is ordinarily effected by the intervention of money. The simple barter thus becomes a complex transaction. There are two exchanges in place of one. There is a sale and a purchase, and then a purchase and a sale. The first vendor employs the purchase-money he has received from his purchaser in the purchase from a third party of the object he desires to use. Our attention is thus directed at each time to one part only of the transaction. We accordingly consider the sale and purchase of a single article and the circumstances affecting the price of that article, assuming for the time that money is an accurate standard of value. In this manner, without the complication of the second set of considerations, the circumstances may be traced which affect the price of any exchangeable object. A man purchases an object because he wants it, and because its attainment involves an effort. His want is personal. His effort admits of delegation, and so of exchange. He pays therefore not for the satisfaction of his want, but for the effort which he has been spared. He does not consider the sacrifice incurred by the vendor. What he regards is not the labour that another has undergone, but the labour that he himself can avoid. Accordingly, price varies with the difficulty of attainment. If that difficulty increase, price will rise; if that difficulty be diminished, price will fall. The range of these variations is between the excess of desirability over difficulty, and the entire absence of the latter. If the element of difficulty be in excess, there can be no exchange; because one of its essential conditions has ceased to exist. If that element disappear, the same result and from the same cause, the absence namely of an essential condition, will ensue. The superior limit then, of price is the point at which the difficulty equals the desirability; its inferior limit is the point at which difficulty disappears.

There are however two parties to every exchange, to both of whom the exchange must be beneficial. It is possible that an exchange may take place which is far from being really beneficial to one of the parties, but which is effected because a partial loss is better than a total loss. A man may from error or from misfortune have misapplied his labour; and he prefers to escape with a reduction in price than to incur an absolute loss. But on whatever terms an isolated exchange at any particular time or place may be effected, a series of exchanges cannot take place, or in other words a trade cannot be established or retained, if the exchange should prove constantly disadvantageous to either of the two parties. One man will not continue to render a service to another, if he thereby entail upon himself a sacrifice greater than the consideration that he receives. The element of difficulty will in his estimation of the transaction exceed the element of desirability; and consequently no exchange can be effected. The price therefore in any given instance is determined by the trouble that the purchaser avoids; but that saving of trouble must on an average of cases coincide with the expenditure of trouble that the vendor incurs. Price therefore constantly tends towards the cost of production.

§ 6. The difficulty of attainment is composed of two elements. One is the actual cost of reproducing the object itself. The other is the number of persons who are prepared to purchase the existing quantity. Where there are more than the two parties, the desirability of the object to the other purchasers presents a difficulty in the attainment of that object to each individual competitor. Difficulty then may be said to depend partly upon the smallness of the supply, and partly upon the largeness of the demand. When both these elements of difficulty concur, the difficulty reaches its greatest height. When they are inverted, the difficulty is at its minimum. When both the supply and the demand are large or are small, the degree of difficulty is intermediate; and may be greater or less according to the circumstances of each case. The element of supply may be controlled by the presence or the removal of some natural obstacle, or of some artificial restriction, or by any of the circumstances which influence production. The demand may be affected by an influx of population, by a caprice of fashion, by any circumstance which induces an unusual number of people to pursue at the same time the same conduct or by any similar cause. But the rise in price consequent upon any such altered demand is not produced by any change either in the properties of the object, or in the purchaser's estimation of its use. Price in such circumstances rises, because each individual finds it more difficult than before to procure the article. In like manner when the supply is reduced, the rise in price is occasioned, not by any change in the intensity or the extension of the demand, but by the increased difficulty of attainment.

The cost of re-production may vary in every degree from a slight obstacle up to absolute impossibility. Those degrees however admit of some classification. The quantity of the object desired either is or is not susceptible of increase. In the latter case, the price will only be checked by the desirability of the object to the purchaser. This is the case of monopoly. Obvious instances are paintings by the old masters, rare coins, first editions, the vocal talents of a prima donna, all which command what is usually called a fancy price, that is a price the limit of which is merely subjective. The opposite case is the ordinary one where the quantity of the object may be indefinitely increased at the same or even a less amount of labour. In this case the limit of price will be objective. The effort requisite for procuring the article will exchange for an equivalent effort in producing it. The difficulty of attainment will be equal to the cost of production.

There may however be more than one cost of production. It may happen that an increase of production is required, but can only be obtained at a greater proportionate cost than the former amount. It may happen that while the quantity required remains unchanged the cost of part, but not of the whole, of that quantity is reduced. In both cases, and for the same reason, there will not be the two prices of the two differing costs, but one price. The purchaser pays for the difficulty of attainment. He has nothing to do with the vendor's labour. That labour may be greater or may be less in any given instance, although its average amount tends towards the difficulty of attainment. Consequently, the increase of the cost increases the difficulty of the attainment not to one person only but to all. The price therefore rises; and the producers at the original cost gain the whole difference between the old, and the new or increased, cost. But on the same principle that a partial increase of cost produces a general rise of price, a partial decrease of cost will leave prices unchanged. In such a case, the difficulty of attainment will remain as before; and the producer at the new cost will gain the entire difference between the new or diminished, and the old, cost. This advantage is generally regarded as extra profit, if the producers personally enjoy it; if they transfer their right to others, the consideration which they receive is termed rent. In a subsequent chapter it will appear how in the one case a desire to share in the advantage gradually brings into general notice the means of the partial cheapness of production, and ultimately tends to reduce the price to the lower or improved cost; and again how the increase of price urges men to seek for improved processes which enable them to neutralize the increased cost which sometimes follows an increased demand.

Prices then within the limit of desirability vary with the difficulty of attainment. In cases of strict monopoly, the price is guided solely by the desire of the purchaser, and has no tendency towards any other limit. In cases of the absence of monopoly, where there is one cost and one price, the price whatever may be its occasional perturbation, steadily tends towards that amount of sacrifice which is involved in rendering the service, and which is called the cost of production. In cases of partial monopoly, where there is a diversity of cost and a uniformity of price, the price in like manner tends to the cost of production of the most costly part. Its residual phenomenon, the difference in amount between the two costs, either augments profit, or appears as rent.

But the cost of reproduction is in some cases modified in a remarkable manner by the character of the demand. As there are degrees in difficulty, so there are degrees in desirability. Men either can or cannot forego the objects of their desire. Some things are absolutely essential to our position in life or to our very existence. With other things, however desirable, we can dispense. These two classes are differently affected by an increase of cost. In the case of superfluities there are always some purchasers for whom the price is at its maximum. In their estimation, the desirability of the object and the difficulty of its attainment are nearly equal. Even a slight increase of cost therefore destroys as far as they are concerned the conditions necessary for exchange. They will cease to satisfy that particular desire. The difficulty will ascend the scale, but not to its full extent. Price will rise, but not to the height that might at first have been expected. There is a counteracting influence at work. The demand of the excluded class was an element of difficulty to other purchasers, and this demand is now withdrawn. The difficulty therefore, although increased at one side, is reduced though not to the same amount on the other. The price will rise to the amount of difficulty caused by the increased cost, less by the amount of difficulty which the suppressed demand represents. In like manner any reduction in cost will bring under the conditions of exchange persons who were previously excluded. Their demand will consequently form a new element of difficulty; and the fall of price will to that extent be checked. The price therefore of superfluities is modified by the diminution or the increase of purchasers. The rise consequent upon the increase of cost is checked by privation. The fall consequent upon a decrease of cost is retarded by enlarged enjoyment.

But the case is otherwise with things that are indispensable. In their case price has no such modifying influence. All men must have food; and no man can use more than a very limited quantity. Hence an increase in the cost of food will not be checked, to the same extent at least as in other cases, by privation. People prefer to sacrifice other sources of enjoyment than to reduce considerably their consumption of food. Accordingly a deficient harvest brings with it a very disproportionate rise in price.* A deficiency of one-third of the crop, even when relieved by foreign supplies, has been sufficient in England to treble the ordinary price. Such a state of things brings with it a general reduction of expenditure in every direction. A famine at once prostrates all industrial exertion with the one sad exception of the trade in food. But the extension of the demand, whatever its intensity may be, is not great. There is no very marked difference in times of abundance and of dearth between the consumption of people in tolerable circumstances. Consequently a plentiful harvest will not bring with it a proportionate increase of consumption. Since people do not require more than a certain quantity of food, the desirability is brought below the difficulty. There is no increase of difficulty arising from increased demand. The expenditure saved is directed towards other sources of enjoyment. Accordingly after a good harvest, trade in its other branches is always brisk, and the price of food falls very low. As the check that controls the rise of price in necessaries is absent, so also is the check that moderates its fall.

§ 7. Both desirability and difficulty are obviously relative terms. They vary according to the circumstances of the person to whom they are applied. Apart from the primary appetites common to the species, both the number and the intensity of human desires are indefinite. The same desires, or the same degrees of desires, are not found in two persons of different nations, or of different generations, or of

---

* Porter's *Progress of the Nation*, p. 429.

different years, or of different occupations, or of different sexes. They are not found even in the same person in different circumstances. In like manner the obstacle which to one man is insuperable is scarcely regarded by another. A porter will carry a weight which a man of ordinary muscular power will hardly move. So, one man will easily satisfy a given desire at a cost which to another man would be ruinous. What may be a reasonable indulgence in one person may be a criminal extravagance in another. What is an absolute necessity in some cases may in others be an idle luxury. But our intense desires are few, and our moderate desires are many; and a large amount of purchasing power, like a large amount of muscular power, is comparatively rare. Hence each increase of difficulty excludes a constantly increasing amount of desire, while every diminution of effort brings larger and larger classes within the necessary conditions of exchange. Every permanent increase therefore of cost rapidly increases the amount of human privation. Every permanent decrease of cost extends in ever widening circles the amount of human enjoyment.

It is but of late years that this principle, obvious as it appears, has been fully recognized. Experience however has now fully confirmed its truth. In those branches of manufacturing industry in which the use of machinery has been most extensive, the demand for labour has so rapidly increased that wages, so far from falling before the competition of invention, have largely risen. In the cotton manufacture for example, where beyond all other cases machinery has been most influential, wages were increased between the years 1839 and 1859 on an average from ten to twenty-five per cent. The reason of this unexpected result is universally admitted to have been the immense increase in the consumption of these goods, which sprung from the reduction in their price. Again, books and newspapers, which if sold at the current prices of thirty years ago would not pay their expenses, now afford when their prices bring them within the reach of numerous readers a liberal return to all who are connected with them. It is now an axiom in finance that the higher the rate of duty upon articles in general use, the less will be the produce. "In no instance is an increase of duty followed by an equal increase of revenue, but on the contrary the produce will be less and less according as the duty advances, until there is no increase of the revenue but a falling off."* In like manner it is stated as the result of all past experience in railway affairs that "there is hardly an exception to the rule that a high fare produces a low amount of traffic, and stints its growth, while a low or moderate fare collects a larger amount of traffic and fosters increase."†

---

* Levi *on Taxation*, p. 63.
† *Journal of Statistical Society*, vol. xxii. p. 295.

CHAPTER XV.

OF THE CIRCUMSTANCES WHICH DETERMINE THE EXTENT OF EXCHANGE.

§ 1. Exchange consists in the voluntary transfer by one person to another person of some means of enjoyment in consideration of a transfer to him or on his behalf by or in behalf of that other person of some other means of enjoyment. It follows therefore that no exchange can take place unless transferable means of enjoyment exist on both sides; and unless these means of enjoyment are of different kinds. The objects of exchange must exist, for otherwise exchange is impossible. They must differ, for otherwise no motive for exchange will arise. Further there must both be two parties to the exchange; and these parties must be persons who are able and willing to deal. Again, as exchange is voluntary and so implies the free discussion of terms, both parties must be placed in communication. An undiscovered purchaser is practically non-existent. Whatever circumstances therefore occasion delay or difficulty in communication, reduce proportionately the number of persons available for the purposes of exchange. Nor is this necessity for communication limited to the mere arrangement of terms. The parties may be able to meet and to conclude their agreement;

but great difficulty may exist in carrying out that agreement, and in effecting the transfer by the delivery of the subjects of exchange. If this difficulty be not absolutely insuperable, it may be so great as to render exchange unprofitable. It thus appears that the power of exchanging, so far as it depends upon natural causes, and is not disturbed by the action of extraneous forces, depends upon and is limited by five classes of circumstances. These are the productive power of the industry of both parties; the variety of their occupations; the willingness of persons to exchange; the facility of finding such persons; and the readiness of access to them when so found.

§ 2. The greater the quantity of enjoyable objects that any man possesses, the greater will be his power of exchanging. Whether he be regarded as a vendor or as a purchaser, his power of sale and his power of purchasing are alike limited by the extent of his possessions. He can, if he be so disposed, exchange all or any of the exchangeable objects that belong to him for any other object that he may desire and with which the owner may be willing to part. But beyond this limit he cannot pass. He cannot procure by any other lawful means, for I do not now speak of gifts, any of these objects save by the transfer of some portion of his property. The wealthier a man is, the better customer he can be. Whatever tends to increase wealth tends proportionately to increase the power of exchanging. Whatever tends to diminish that wealth, tends in the same degree to diminish that power. In every such increase of production there is a twofold gain, and in every such decrease there is a twofold loss. The account must be taken not on the actual product as in the hands of the producer, but upon that produce as increased by exchange. If the original product represent the labour of one hundred hours, and if by exchange it is made to represent the similar labour of one hundred and twenty hours, the loss upon the reduction of that product by one-fifth will be the product not of twenty hours, but of twenty-four. In like manner a similar increase would give as the gross product in exchange not one hundred and forty, but one hundred and forty-four.

§ 3. In considering the productive power of industry as a limit to the power of exchanging, the direction which that industry may have taken does not seem to be material. Provision must indeed be first made for the necessities of existence before we can take any thought for its refinements and it has therefore been urged that the productive power of agriculture in any community is the limit of that community's power of exchange. If we take into account the whole world, and assert that the number of disposable hands for other employments depends upon the productive power of the industry of those who are engaged in raising food, the proposition can hardly be disputed. In a more limited point of view however, both individuals and nations may, and actually do, purchase their food; and thus the small amount of their agricultural industry, so far from limiting their power of exchange, actually necessitates an amount of exchange which, if they had supplied themselves with food, would have been needless.

A man who supplies all his own wants has no need of exchange. If after providing for his primary wants he have a surplus and can find a market, he can exchange to the amount of that surplus: and so his power of exchanging is limited by the productive power of his industry as applied to the production of food. If he devote his time to the preparation of some other product, and obtain his food in the ordinary way by a preliminary sale and a subsequent purchase, his exchanges are limited only by the results of his general industry, to whatever point that industry may be directed. There are examples even of whole countries thus procuring their food in exchange for their own peculiar products. Although we see instances of it in every case in which any quantity of food is imported, we have some cases on a larger scale. What continually happens between different parts of the same country, occasionally happens between different countries. Holland and Genoa in their most flourishing days never attempted to raise their own food. The people of Barbadoes obtain nearly all their provisions from the United States, and confine themselves to the production of sugar and molasses. For several years after the first discovery of the goldfields, Victoria imported all her breadstuffs, and produced merely gold and wool. In these cases the purchasing power of Barbadoes and of Victoria is limited not by their agricultural but by their general industry. Thus the amount of food in any nation, and consequently the amount of its population, do not necessarily depend upon the quantity of breadstuffs that it is itself able to raise, but upon the quantity that it can from any source whatever procure. In other words, its expansion is controlled not by the productivity of its agriculture alone, but by the productivity of its general industry. In the present state of the world, and during a future so indefinite that we need not concern ourselves as to its close, food like all other commodities can be readily purchased when the means of purchasing it exist. In the commerce between nations, as in the commerce between individuals, it is much more frequently the means to purchase than the object to purchase that men want.

Although these principles are sufficiently obvious in the case of individuals, their application is less frequent, and its clearness is more obscured in the case of nations. In the

s

backward state of the political arts that even still prevails, the national policy of most countries has almost invariably been adverse to exchange. Attempts have constantly been made to isolate nations, and to render each community as far as it was possible self-supporting. In such circumstances when the natural tendency to international exchange is disturbed, when every effort is used to turn the capital and the labour of the nation from other industrial pursuits to the raising of food, or when the relations with other countries are not sufficiently developed to admit of commercial transactions, the power of exchanging is in fact limited not by the general industry of the nation but by that surplus of its industry which remains after it has raised its own food. This case indeed hardly furnishes an exception to the general principle. The food raised in the country is in that particular state of society the only food which the nation is able to procure.

§ 3. The second circumstance which affects the power of exchanging is the variety of the objects of interchange. These objects may vary either from the difference of the natural agents that are at the disposal of the parties, or from the different directions that from whatever cause their industry assumes. When the productive powers of the country are diversified, and when the spontaneous distribution of industry is unaffected by any disturbing forces, there will be ample material for exchange. When on the contrary the products of a country are uniform, if from that or any other cause the industry of the country be confined within few channels, exchange can find little place. Thus in France where the vine flourishes in one part of the country but is unsuccessful in another, the vignerons of the South find their natural customers in their countrymen of the North. In Russia, the Forests, the Ural district, the region of the black soil, and the Steppes, form four great natural and mutually dependent divisions. None of them separately possesses the conditions of complete independence; and their commercial relation secures and consolidates their political unity.* In America the South used to exchange its tropical products for the coal and the iron and the manufactures of New England; and both alike dealt for their breadstuffs with the West. In England the specific characters of the great centres of industry are still more clearly marked; and the mutual dependence of the districts is consequently still more complete. In other countries however, the case is very different. In India the country has been from time immemorial divided into village communities. These little republics, as they have been called,† are almost exclusively agricultural; and contain nearly everything they can want within themselves; and are almost independent of any foreign relations. To this form of government the people cling with surprising tenacity; and as each village supplies its own wants, and as their wants are similar, and the produce of the country uniform, they have no need of mutual assistance. In Italy in like manner a similar municipal form, although on a somewhat larger scale, seems to have been equally prevalent. The population is for the most part concentrated in towns, and in the district immediately adjacent. Each district produces the same kinds of produce, corn, wine, oil, silk, and fruits. "Each city or town," says Mr. Laing,‡ "within its own circle suffices for itself, is a Metayer family upon a great scale living upon its own farm, and having no dependence upon or connection with the industry interests prosperity or business of its neighbours in the land, and very

---

* Haxthausen's *Russian Empire*, vol. i. p. 20.
† Elphinstone's *History of India*, p. 64.
‡ *Notes of a Traveller*, p. 271.

little communication or traffic with any other masses of population by carriers waggons carts diligences, or water conveyances, the objects of interchange being from the general bounty of nature but very few between them." The same acute observer* has noticed a similar state of facts in Germany. " From the shores of the Baltic to the banks of the Lake of Constance, rye, wheat, barley, flax, sheep, cattle, wood for fuel and house building, and all the other primary necessaries of life, are produced in each district in sufficient abundance for its own consumption. It could exist if all the rest of Germany were submerged five fathom beneath the sea. Nay, every agricultural family in the district, and the forty millions are chiefly an agricultural population, stand in the same isolated, unconnected, social condition, producing all it consumes, making its own clothing, preparing its own fuel, and scarcely requiring to exchange industry for industry, even with the smith, the carpenter, the baker, the butcher, or shopkeeper ; living in short in juxtaposition, rather than in social relation with any other family."

§ 4. A further limitation to the power of exchanging consists in the mutual disposition of the parties. They must not only possess objects of reciprocal desire, but they must be willing and ready to transact business. This limitation is at the present day more apparent in international than in private relations. We seldom see, where gain is the sole object, the interference of merely personal feeling: and when in commercial affairs such feelings do occur, their effects are lost amid the number of persons who are not under their influence, and are ready at any time to buy and sell on the usual terms. It is therefore when we have to deal with national policy rather than with the internal transactions of

---

* *Denmark*, p. 11.

any community that the reluctance to exchange attracts our attention. Instances of its existence occur in the rigorous alien laws of the Spartans; in the far-sighted policy by which the Roman Senate sought to weaken the cohesion of conquered federations; in the semi-instinctive repulsion of the South Sea Islands; and in the traditional seclusion of China and Japan. Where no actual prohibition exists, the conditions on which foreigners are admitted are sometimes so onerous as practically to exclude them. In France the Crown claimed the right of succession to the deceased alien. At the present day no British mercantile establishment has ever been formed in Porto Rico* because every foreigner who desires to settle in the island must make a public profession of the Roman Catholic religion, and must swear allegiance to the Crown of Spain. According to the old law of England, it is said that alien merchants could come but four times in the year and remain but forty days on each occasion.† Political writers have often noticed with admiration that chapter of the Great Charter which guaranteed to foreign merchants the right to enter and leave the realm at their pleasure, to remain in it without limitation of time, to travel unimpeded throughout every part, and to buy and sell exempt from all "evil tolls," and subject only to the ancient and allowed customs. Unfortunately English legislation did not always adhere to this enlightened policy. Without noticing the various statutes that have been passed against aliens, it is enough to refer to the acts that declared the trade with Ireland and the trade with Scotland to be nuisances, and the long list of prohibitory duties on foreign commodities. There is indeed little difference between the restrictive laws of modern European and American policy

---

* Levi's *Annals of Legislation*, vol. vi. p. 17.
† See *Second Institute*, p. 57.

and the rude xenelasy of the Spartans, or the selfcomplacent exclusive system of the Chinese. There may doubtless be cases in which such churlishness becomes necessary. It would be small wisdom to supply an enemy on any terms with the munitions of war. It was no impeachment of the liberality of the British Government that during the Crimean war they would not allow British capitalists to contribute to the Russian loan, or British manufacturers to execute Russian contracts. But these are cases in which for other reasons men knowingly submit to a loss, or at least fail to realize a gain; and they have nothing in common with those cases where from a mistaken notion of the true nature of exchange all peaceful intercourse is prohibited, and a barbarous isolation is habitually maintained.

§ 5. The productivity of industry, the variety of employments, and the willingness to trade, are thus all included in the conception of a purchaser. A purchaser is a person who possesses some object that we have not but desire to have, and who is willing to negotiate an exchange with us. But it is not enough that such persons should somewhere exist, if they cannot be found when we require them; or if when found they cannot be reached. We may possess exchangeable objects with which we desire to part; and we may be satisfied that persons are in existence who possess other exchangeable objects suited to our wants and are as desirous as we are of an interchange; but if we cannot find such persons, we can effect no exchange. There is thus a further limitation to the power of exchanging in the possibility of finding purchasers. The force of this limitation is attested by the number of the contrivances, to which I shall presently refer, that society has adopted for diminishing the impediments to intercourse. But we see daily instances of its operation in the quantities of unsaleable goods that from time to time glut our warehouses and our stores. It is not that there are no persons in the world who would purchase these goods at a remunerative price; but that it is impossible to find such persons. So in like manner many professional men of undoubted ability waste their lives without business, not because they have nothing to sell or are unwilling to deal, or that the public is unable or unwilling to employ them, but because they are unknown. On the other side many men, whose attainments are frequently of a very moderate kind, reach wealth and eminence from no other cause than that accidental circumstances or their own skill in advertising have made the public familiar with their names. We can hardly however appreciate the full importance of this condition. In the absence of any artificial arrangements the difficulty of finding a purchaser is far greater than we with all our familiar appliances of exchange can easily understand. For it is necessary that the trader should find not only a purchaser who desires, having reference both to quantity and quality, the object that the vendor has to exchange; but also a purchaser who has in like manner such an object of exchange of such quality and in such quantity as the vendor requires.

§ 6. Connected with the last subject, though distinct from it, is the facility of access to market. It is one thing to ascertain where a customer can be found: it is another thing to be able to reach him. It is certain that there are purchasers in Central China who are able and not unwilling to deal with Englishmen; but the difficulties of access are still insuperable. Lord Durham in writing of the settlers in the remote parts of Western Canada observes that they have "a rude and comfortless plenty but no wealth." These men occupy a rich and virgin soil, and receive from the bounty of nature a return for their labour far more than sufficient for their personal consumption; and they fully appreciate the advantages of exchange. But the population is scattered; and each man is engaged in nearly the same occupation as his neighbour. Hence no home market exists among them; while the absence of roads and all other means of communication cuts them off from all access to the markets of more populous districts. Although therefore they have a surplus produce, different from that of other people, although they are willing to trade, and know where customers can be found, as they have no means of reaching the market, they are unable to exchange. In like manner it not unfrequently happens, in districts where roads have not been formed, that fruit or potatoes or other produce, when the harvest is unusually abundant, are allowed to rot in the ground, because in the existing circumstances the cost of carriage would not be compensated by the market price. The difficulty of access to a market which the cost of carriage represents, depends partly upon the bulk of the commodity, partly upon the character of the country over which it must pass, and partly upon the means available for locomotion. When the commodity is difficult to move, when the space to be traversed is great or difficult, when the means of locomotion are scanty and imperfect, the effort to overcome these obstacles will be so great that exchange will become burdensome and will therefore cease. As those obstacles or any of them are reduced, the motives to exchange will have room to operate, and exchange will be proportionately extended. The importance of this access to a market is shown by two admitted facts. Of these the first is that by far the greater part of the price of almost all commodities goes to the distributor and not to the producer. It has been estimated that the cost of distributing the produce of the soil is double the rental of the soil.* The other fact is that men have found it profitable to expend and will probably continue to spend vast sums upon the construction of railways, roads, and shipping. The whole of the latter and no inconsiderable part of the former charge is rendered necessary by the natural obstacles that exist to the transport of commodities.

§ 7. The extent to which exchange will spontaneously proceed is regulated by the same principle as that which determines the natural limits of the other auxiliaries to industry. Like them, exchange will continue so long, and so long only, as it fulfils its end. While its practice is felt to aid industry, it will continue. When it ceases to give such aid, much more when it becomes not an aid but a burden, not a source of gain but a source of loss, it will cease. It ceases to give the aid which it was designed to afford, when one of the parties either is able otherwise to obtain with equal advantage the service which he purchases, or prefers to forego that particular gratification than to pay for it the price demanded. In the absence therefore of any disturbing cause, the cessation of exchange is evidence that in the circumstances it has ceased to be profitable; while its continuance is in like manner evidence that the profit continues. In the former case its revival at best brings no gain, and will probably cause loss. In the latter case, by the very terms of the question, it secures a greater amount of enjoyment at a less sacrifice of time and trouble than could otherwise be obtained. It follows therefore that where it is proposed to interfere with the natural development of exchange, those who propose such a measure are bound to prove its expediency; and also that this proof must be supported by other than pecuniary considerations. Impediments to the development of exchange

---

* *Journal of Statistical Society*, vol. xxii. p. 414.

can never increase but must always diminish wealth. They imply a pecuniary loss, although it may be expedient on sufficient reasons to submit to that loss. Those persons therefore are not to be heard who propose by any such interference to promote a country's wealth. The advocates of interference are bound to admit the loss, and to show the presence of some compensating advantage. It was on this ground that Smith, and, long before him, Bacon justified the policy of the Navigation Laws. It is now generally conceded that their argument was founded in error; but on the assumption of its truth their plea was valid. "The king also for the better maintenance of the navy ordained that wines and woods from the ports of Gascoign and Languedoc should not be brought but in English bottoms; bowing the ancient policy of the estate from consideration of plenty to consideration of power : for that almost all the ancient statutes invite (by all means) merchants strangers to bring in all sorts of commodities, having for end cheapness, and not looking to the point of state concerning the naval power."*

---

* Bacon's *History of King Henry VII.*, Works (Spedding's edition),vol. vi. p. 95.

CHAPTER XVI.

OF THE RECIPROCAL INFLUENCE OF THE INDUSTRIAL AIDS.

§ 1. The operation of these aids is not confined to the direct assistance that they severally afford to industry. They also largely influence each other. Each of them is thus most powerful when the others are most developed; and consequently every improvement in the development of any one of them tends to promote the efficiency of the rest. This reciprocal influence forms the subject of the present chapter.

In no case is this influence more clearly seen than in that of capital. Besides the direct services which capital renders, its formation is a condition precedent to the extended use of all the other aids. The necessity for the assistance of capital in the process of invention has been already shown. Knowledge, both theoretical and practical, and dexterity are essential to invention. But knowledge implies leisure; and dexterity implies practice; and both leisure and practice imply in turn the means of subsistence. Besides, inventions are seldom produced in a complete state. They are generally the growth of time: they involve many partial, and many complete failures; and they are the subjects of continual improvement.

In most cases the cost of the instruments through which the invention has to operate is very considerable. A solitary man therefore or a small and poor community can never be very inventive. It is only where the inventor can obtain the aid both of capital and of co-operation that he can with any reasonable hope of success attempt to extort her more important secrets from nature. In a state of isolation, no such aid is at all procurable. In a poor community, it may to some small extent be obtained. In any case the inventor himself has seldom the means of carrying out his projects. His pursuits are not conducive to wealth. The time and the absorbing attention which any great invention demand, furnish very little opportunity for present gain. Where people are poor, or where the want as compared with the circumstances of the population is not very urgently pressing, few persons will consent to risk any considerable sum on the untried projects of a dreamer. Even at the present day almost every important patent is worked by the aid of some capitalist. But if an example be sought of the different rates at which inventive talent grows where capital is abundant and where it is scarce, it is sufficient to compare the England of the nineteenth with the England of the eighteenth century. At the present day the services of the engineer and of the chemist are liberally rewarded; and every invention that offers any reasonable prospect of success quickly finds support. In the days of George II., the case was very different. Works of acknowledged utility were put aside from want of funds. The most fertile lands of Lincolnshire were left under water because the county was unable to raise the enormous sum of £40,000.* Most of the early Acts of Parliament for the improvement of the navigable rivers, modest as these projects usually were, re-

---

* Smiles's *Lives of the Engineers*, vol. ii. p. 51.

mained from want of funds inoperative.* The profession of engineers had no existence. Some millrights were sufficient to supply the wants of the country. In such circumstances there could not be, and there was not, any class of inventors. A few men of rare genius overcame every difficulty and laid the foundations for the mechanical triumphs of the present day. But while these great men gave the impulse, the numerous minor improvements that are almost daily made in every branch of constructive art are the work of practitioners, to whose talents the extension of these arts, itself created by and in its turn creating capital, has given adequate scope.

§ 2. Although when compared with the unconnected efforts of individuals a great economy of capital is effected by co-operation, yet where the associated industry is at all extensive the absolute amount of capital is very considerable. The workers must be maintained until the product is finished. No small portion of the influence of co-operation depends upon the skill to which it gives rise: but the exercise of this skill depends upon the improved instruments that it employs; and these instruments are generally costly. There must also be an adequate and constant supply of the materials which these workers with the aid of these instruments are to convert into finished products. But as the produce of their joint efforts is far greater than the produce of the sum of their separate efforts, so a number of men working in concert require a larger supply of material than the same number working separately. If eighteen unconnected workmen make in a day 360 pins, or about one ounce in weight, they will only require to have in store for each day one ounce of metal. But if the same men by

---

* Ib., vol. i. p. 303.

means of co-operation make upwards of 86,000 pins, the daily supply of metal which they will then require will be 240 ounces. If we assume a month for the replacement of this part of the capital, the proprietor must have at all times upon hand in different stages of progressive manufacture thirty times 240 ounces or about 450 lbs. of metal.

The annual amount of wages paid in Great Britain has been estimated at two hundred millions of pounds sterling. This sum is spent in the remuneration of labour exclusively, and does not represent the enormous sums that have been sunk in machinery, in buildings, the means of transport, and other appliances of industry; or that are spent in the purchase of raw materials. If these advances were withdrawn, if each labourer had while his work was still unfinished to provide for his own immediate wants, and also to procure his own implements and his own materials, there can be no doubt that, however strong might be the desire for co-operation, or however clear the perception of its advantages, its actual practice would not be very extensive.

§ 3. When the production of a commodity is complete, if the producer do not design it for his personal use, he must find for it a market; he must convey it to that market; and, if the state of the market be unsatisfactory, he must hold it over until he can effect a sale on advantageous terms. All these proceedings involve both time and trouble; and consequently imply the command of some accumulated funds. But this statement only faintly represents the dependence of exchange upon capital. The assistance which capital renders in this respect is now so familiar that it is almost as little heeded as is the unimpeded circulation of the blood. Yet a very small amount of attention is sufficient to show that exchange can never be extensive in a poor community. So onerous are these incidents of exchange that numerous expedients have been devised to reduce their cost. Upon the extent to which these expedients are adopted and the success with which they are used, the power of exchanging mainly depends. But all of them are more or less costly; and thus the means of defraying that cost are a condition precedent to the extension of exchange. It needs but to remember the fabulous sums that are invested in facilitating transport in England; the capital represented in its roads, its railways, its canals, and its shipping; the expense of the establishment and of the maintenance of the currency; the vast system of retail dealers in all their variety; and the whole wonderful apparatus of circulation in the social organism. When we think of these things, we shall have little difficulty in comprehending the extent of the obligations that capital has conferred upon exchange.

§ 4. Since capital includes food and materials, every reduction in the period of labour effects a corresponding economy in the means required for the labourer's support; and every reduction in the quantity of materials consumed in the process is by the very terms an economy of material. The introduction therefore of natural forces in lieu of or in addition to human powers sets free a quantity of commodities which were previously expended in the support of the labourer, or in materials now become superfluous. In the cutting of wood and of precious stones a large saving is obtained by the use of proper tools; and where machinery has been applied this saving is still more conspicuous. In the process of bleaching, as we have already seen, the use of proper chemical agencies has both effected a great saving of time, and set free for other uses a considerable extent of land. Careful experiments have shown that in the old method of inking type with hand by balls, and in the new method of performing the same operation by machinery, the consumption of ink by the machine was to the consumption by the balls in the proportion of four to nine.* Another instance of this economy is found in the application of manures. "The common drills says Mr. Pusey,† economise manure by concentrating it in lines along the rows of the turnip plants. Thus instead of shovelling bones from carts as was first done in Lincolnshire at sixty bushels per acre, we came to sow only sixteen bushels of bones in lines, or more recently but three bushels of super-phosphates prepared either from bones or from the animal remains of geological ages, where Liebig told us, and told us truly, to search for our phosphorus. But though turnips are sown in lines and come up quickly in lines, no sooner are the thriving young plants well marshalled in green array than nineteen in twenty are ruthlessly cut down by the hoe, so that the field appears for a time once more bare. The roots must of course be allowed ample room in the row, but some manure will have been wasted in nourishing the plants doomed to perish. Hence Mr. Hornby's drop drill avoiding this wholesale massacre is made to drop the seed and the manure by a second step of mechanic frugality only at those points in the line where the plants are intended to stand. Such is the elastic yet accurate pliability with which in agriculture mechanism has seconded chemistry."

§ 5. The influence of invention upon co-operation seems at first sight inconsiderable. The combination of human action affords little room for the introduction of physical agencies. Yet the extension of invention tends to produce one form at least of co-operation. The perfect implement requires the skilful worker. The skill of the worker depends upon the division of employments. Where a

---

* Babbage's *Economy of Manufactures*, p. 46.
† *Jurors' Reports*, 1851, p. 230.

man possesses an aptitude for the use of any unusual implement, his powers are wasted if he leave that implement and engage in any ordinary occupation. The practice of its use soon gives increased skill; and thus the speciality of the business becomes continually more marked. There is accordingly no important invention which has not given rise to a multitude of distinct employments. The steam-engine has brought into existence various classes of mechanical engineers and of engine drivers and other subsidiary officials. Our fathers knew nothing of electricians and telegraphic operators. The photographer, and lithographer, and the photolithographer, are all now recognized artists, as much as the chemist, or the engraver, or the printer. Even that occupation which was once almost synonymous with stupidity, that of the ploughman, is fast losing its degrading associations. Its character has been elevated by the improvement in the instrument.* The efficient use of the modern plough requires and promotes no inconsiderable intellectual development. It is no longer a task that may safely be entrusted to any clown. The invention requires a skilled workman; and a skilled workman implies a special occupation. In the manufacture of cotton, in the manufacture of wool, in short in all the great branches of industry, the use of invention has always been attended with a more elaborate organization of labour. Thus the progress of invention and the progress of co-operation are mutually dependent. The success of the invention depends upon the extent of the co-operation; the extent of the co-operation is limited by the power brought into action by the invention.

§ 6. The interchange of their respective possessions by two persons does not admit of the substitution of any physi-

---

* See Mr. Pusey's *Report*.

cal power for any part of the human agency. But among the circumstances which regulate the power of exchanging are the means of communication, as well for the discovery of persons willing to exchange, as for convenient access to them. Some of the most important attempts that have been made to facilitate such communications belong to the class of inventions. The inventions indeed that have been contrived for this purpose form the great triumphs of modern art. The telegraph and the steam-engine have increased the means of communication between the exchanging parties: and the steam-engine has given unequalled facilities for the delivery of commodities. The extent to which exchange has been stimulated by these great inventions cannot easily be described. The reduction of the cost of transit, and the more accurate information now promptly obtainable, lower the general cost of production, and so diminish the difficulty of attainment. Many persons are consequently brought within the conditions of exchange that previously were excluded. In Wiltshire the annual consumption of artificial manures used never to exceed 150 tons. Since the extension of the railway system to that county, it amounts to between 3000 and 4000 tons. In many districts the price of coal has been reduced from thirty to forty per cent.; and the purposes to which it has been applied have consequently been largely increased. The extraordinary development of London, and the amount of exchanging that this development implies, were rendered possible by steam. The graziers of Aberdeen and of Devon, the gardeners of the western counties, the dairymen of Somersetshire and of Lincoln, now find in the great metropolis a ready market for their respective produce. The extension which steam has thus given to exchange enables three millions of people in a district absolutely unproductive of food, without fear or inconvenience, to dispense with the keeping in stock even a fortnight's supply of provisions. Modern London, it has been truly observed, is mainly fed by steam. The express meat-train runs nightly from Aberdeen to London, and makes the journey in twenty-four hours. Express fish-trains from Dunbar and Eyemouth and various towns on the coasts of Northumberland and of York arrive in London every morning. "And what with steam vessels bearing cattle, meat, and fish, arriving by sea, and canal boats laden with potatoes from inland, and railway vans laden with butter and milk drawn from a wide circuit of country, and road vans piled high with vegetables within easy drive of Covent Garden, the Great Mouth is thus from day to day regularly, satisfactorily, and expeditiously filled."[*] In the fish trade the influence of steam is especially conspicuous. In the inland districts of England fresh sea-fish were formerly almost unknown. At the present day such fish is used by almost every family of the middle class in every considerable town. More fish are now carried from the east coast to London in a fortnight than in old times were carried thither in a year.[†]

§ 7. The influence which co-operation exerts upon capital resembles in many respects its influence upon labour. It consists partly in the effective addition to the amount of capital, and partly in the economy of its application. Many small accumulations, although separately insufficient for any industrial purposes may, when combined form a very considerable sum. There are many persons who possess a large capital, but will not venture the whole of their property, although they are willing to incur a moderate risk. Again there are many undertakings which either for their

---

[*] Smiles's *Lives of the Engineers*, vol. i. p. 191.
[†] *Edinburgh Review*, vol. cvii. p. 401.

execution or for their regular and efficient management would exceed the resources of even the most wealthy individual. There are others of which the success depends upon the amount of public confidence that they enjoy; and this confidence can only be acquired by the reputation derived from the security which a numerous and wealthy proprietary affords. Thus the collective contribution of capitalists, like the combination of men's physical powers, both utilizes the forces which would separately be inefficient, and accumulates in the proper time and at the proper place the amount of force required for any undertaking. But this is not the whole amount of the advantage derived from the co-operation of capital. Capitalists may co-operate not only with other capitalists but with labourers. As in its former aspect this form of co-operation is analogous to the combination of labour, so in its latter aspect it presents a resemblance to the division of employment. Co-operation collects the scattered, and vivifies the idle, capital; and it also tends to place the capital so collected and so vivified in the hands most capable of advantageously using it. It is a common practice in mercantile life to advance goods to some person in whose abilities and integrity the capitalist has confidence. Many eminent men have owed to this practice their success; and as intelligence is diffused and the standard of commercial morality raised, this mode of mutual service will become more frequent.

The most convenient illustration of both these results, the aggregation of small sums, and the proper distribution of the collective capital, is found in the operations of banking. This great system which has gradually grown out of the wants of the public both to keep money safely and to obtain its temporary use, belongs indeed to a very late stage of society; but its principle is sufficiently simple. The bank collects for the purpose of safe keeping and on condition of immediate repayment on demand, from a great number of customers sums of varying amount but in the aggregate very large. Experience has taught the average amount which the owners daily claim; and the balance after provision has been made for these demands is consequently available for the use of a different class of customers, those namely who desire to borrow. Since the bank is responsible to its depositors for the money, it is careful not to lend to any persons but those whose circumstances or character hold out a reasonable prospect of success in their undertakings. Constant practice and the ample means of inquiry which a bank possesses enable the managers to form a tolerably correct judgment of their customers. Thus by the receipt of deposits, small capitals are collected into adequate sums, and idle capitals are rendered productive; and by the operations of discount, these sums find their way into the hands of persons who are in a position to use them more efficiently than their owners could do.

It is obvious that co-operation in proportion as it increases production increases the fund from which capital can be saved. This effect it has in common with invention, and indeed with every circumstance which tends to the increase of wealth. But besides this increase to the possibility of saving and to the efficiency of the sum so saved, the division of employment economizes the amount of capital required. I have already observed that the division of employment saves not merely the time of the labourer in his apprenticeship, but also the capital which is then required for his support and for the material which is wasted during his first unsuccessful essays. But there is a further economy in the retention by each labourer of no greater number of implements than he can keep in constant operation. The practice of the various arts implies the existence of a great variety of costly tools by which these arts are carried on.

But as a man can only exercise one art at the one time, the greater the number of the arts which he professed to exercise, the greater would be the amount of his capital constantly remaining idle. If a man possessed complete sets of tools for six different trades, five of these six sets must always be unemployed. There is therefore a clear gain, and that in the aggregate of no small amount, when each individual is enabled to limit his requirements to that single set of implements which he can keep in continual use. "The advantages of such a change to the whole community, and therefore to every individual in it are very great. In the first place the various implements being in constant employment yield a better return to what has been laid out in procuring them. In consequence their owners can afford to have them of better quality and more complete construction. The results of both events is that a larger provision is made for the future wants of the whole society."*

§ 8. One of the advantages that Adam Smith ascribed to the division of employments is the impulse that this practice gives to invention. This result is indeed rather a consequence of the increase of skill that co-operation brings with it than a direct influence upon the inventive process. In either case it is too contingent and too remote to have led, even if it could have been anticipated, to the introduction or even to any considerable extension of separate occupations. When a person has become familiar with the processes of any subject, he possesses at least one essential qualification for suggesting improved methods of performing those processes; and such familiarity is generally attained by that exclusive attention to which the division of employments gives rise. It is therefore reasonable to expect that

---

* *Rae's New Principles of Political Economy*, p. 164.

practical improvements in any art will come from the professors of that art; and the tendency towards such improvements will be the more conspicuous in proportion as the art becomes more special. Yet it may well be doubted if this tendency deserve all the influence that Adam Smith has attributed to it. Nothing is more striking in the history of modern inventions than the unexpected quarters from which important novelties are constantly derived. The founder of the cotton manufacture was a barber; the inventor of the power-loom was a clergyman. A farmer devised the application of the screw propeller. A fancy-goods shopkeeper is one of the most enterprising experimentalists in agriculture. The most remarkable architectural design of our day has been furnished by a gardener. The first person who supplied London with water was a goldsmith. The first extensive maker of English roads was a blind man, bred to no trade.* The father of English inland navigation was a duke, and his engineer was a millwright. The first great builder of iron bridges was a stone mason; and the greatest railway engineer commenced life as a colliery engineman. Practical knowledge is not the sole or the most important condition of inventive success. It is true that in complex subjects changes can be safely effected only by those who are acquainted with all the details of the subject and their mutual relations; but the mere familiarity with the routine of a process which frequently constitutes the sum of the practitioner's attainments, is very different from an accurate knowledge of its principles. Accordingly persons whose usual occupations are wholly dissimilar often obtain a theoretical acquaintance with the processes of other businesses sufficient to enable them, if not to practice the art, at least thoroughly to comprehend its details. Such persons too are

---

* Smiles's *Lives of the Engineers*, vol. i. p. 207.

less subject to disturbing influences than practitioners. They are less embarrassed by the traditions of the art. They are not restrained by the natural reluctance to confess in their declining years that the operations in which they prided themselves and all the hardly won skill of their youthful days are at best worthless. Thus as their stand-point is different, they escape from the common confusion between the utility of a process and the familiarity with its execution.

§ 9. Whether exchange gives rise to the division of employments, or whether the division of employments determines exchange, is, says M. Bastiat, a subtle and idle question. No such question could at all arise if the true position of these agencies were rightly understood. Each of them is an independent aid to labour; but each re-acts upon and modifies the other. The incident of exchange which renders it dependent upon the division of employments is the necessary diversity of its objects. This diversity is evidently increased as the various classes of occupations become separated. Such a separation is indeed limited by the extent of the market; but unless it take place, no market to any considerable extent can at all exist. Without distinct occupations the desirable objects would be for the most part of the same kind; and their number would be very restricted. In such circumstances there would be little room for exchange. Accordingly we find the least amount of industrial activity in those countries in which the industry exhibits the closest approach to uniformity.

In Turkey where there has been long and gross misgovernment, and where the absence of roads prevents the possibility of finding in foreign trade the market which misrule has checked at home, three-fourths of the population are engaged in the same agricultural pursuits. In each district the agricultural classes generally cultivate the same articles of produce, and follow the same routine of culture. Every man therefore may possess a superfluity; but it is only of those very objects which his neighbor, so far from wishing to buy, is desirous to sell. No exchange can consequently be effected; and production is therefore kept as low as the fertility of the soil will permit. Throughout the whole country the same monotonous existence prevails, and the same absence of all attempts at improvement. In France a similar result has been to some extent brought on by the law of the compulsory division of inheritances. The tendency of this law is to divide into unnaturally small portions agricultural industry, and to throw an unnaturally great proportion of the population into the same occupation. The people for the most part supply their own few wants. The town which furnishes them with such commodities as they cannot themselves produce has but a small amount of business to transact. The country is poor, and the town which it supports is poor. There is an unnatural development of a particular class; and the result is injurious as well to the part itself as to the entire organism. The same result, although proceeding from somewhat different causes, has hitherto existed in Ireland. In that country, partly from the repressive commercial policy pursued in former days by England, and partly from the state of the law which practically prevented all commercial dealings with land, the population was chiefly engaged in the production of their own food. In the more remote districts every man was like his neighbor. There was little variety of employment, and consequently little exchange. Hence it was that when the famine came the distress was universal. The single occupation, that of growing potatoes, on which the bulk of the population depended, was gone. There was no other business which might serve to break the violence of the calamity; and there existed among men who lived in commercial isolation no distributive agency to deal with the supplies of food that were collected from other countries.

In strong contrast with these cases we may observe the state of England. In no other country is exchange so frequent. In no other country are employments so varied. In the cotton manufactures of Lancashire, the number of occupations returned by the people themselves, although perhaps somewhat excessive, amounted in 1841 to upwards of 1200. In the census of 1851, the occupations of the people as described by themselves amount to several thousands; and although their descriptions are often inaccurate, it would seem that the actual number of occupations, whatever may be the case as to the number of persons employed in each branch, is not materially affected. The census commissioners enumerate 1057 groups of occupations for males and 746 for females, although many of the latter class coincide, either strictly or with but little deviation, with those assigned to males. Throughout the present century there has been a steady decrease in the per centage of the population whose labour was required for the production of food ; and a steady increase in the per centage of those who are engaged in other employments. If the three parts of the United Kingdom be compared, there will appear a remarkable augmentation in the number of the employments in the richer as compared with the poorer. According to the tables given by Mr. Spackman in his work on the occupations of the people, it appears that out of a list of 785 occupations relating to manufactures, trade, and commerce, 740 are pursued in England; 501 in Scotland; and but 261 in Ireland. We see therefore both the proximate cause of the vast amount of exchange of which England is the centre ; and the superior complexity of social structure which characterizes the wealthier country.

§ 10. When in any society the division of employments has been established, each producer confines himself to one kind or at most to a few kinds of industry, and relies for the supply of his other wants upon his power to obtain the appropriate enjoyable objects in exchange for the results of his own industry. His command therefore over " the necessaries, conveniencies, and amusements of life " is regulated by the quantity of exchangeable objects that he possesses, and by the facilities at his disposal for effecting their exchange. Both these conditions are essential, since in every exchange a man must have something to give, and must find something to get. Assuming consequently the quantity of objects for sale at the disposal of any vendor to be constant, his opulence or his poverty must depend upon his power of exchanging these objects. The person therefore who can sell the whole produce of his industry as quickly as he produces it, is richer than the person who with equal appliances is obliged to retain for some time his produce before a purchaser appears. If these two persons could produce their respective commodities in six months, and if the former sold his goods immediately and the latter sold his goods at the end of a second period of six months, the result would be that the person who had a ready market would have produced exactly twice as much at the end of the year as his less favoured competitor. In practice, very few persons have an extent of business commensurate with their capital, or turn over their capital as rapidly as its nature would permit. Notwithstanding all the expedients to facilitate exchange which the ingenuity of man has contrived, large masses of capital, even in these communities where exchange is most extensively practised, are always partially idle; and so fail to perform their proper industrial function. Of the importance of this principle of quick returns Mr. Mill* has pointed out three signal proofs.

---

* *Essays*, p. 56.

One is the large sum often given for the goodwill of a particular business. Another is the large rent which is paid for shops in thoroughfares and similar situations. The third is the fact that in many trades there are some dealers who sell articles of an equal quality at a lower price than other dealers. The principle on which such persons avowedly conduct their business is that by keeping the whole of their capital in more steady employment, they obtain a larger gain upon the whole of their transactions, although on each particular operation they have a smaller profit than their competitors. Since capital limits industry, the quickening of its action not merely enables the same amount of industry to be conducted, and the same organization to be obtained, with less capital and therefore at less cost than would otherwise be necessary; but also it calls into increased activity both labour and natural agents, and so leads to increased production and consequently to a better industrial organization. Thus a brisk market aids capital by enlarging its field of employment. It stimulates the investment as capital of any accumulations that his forbearance or his credit have placed at the labourer's disposal; and so by fulfilling the necessary conditions utilizes to the utmost that labour and those natural agents which in the absence of the assistance of capital had remained idle or had been only partially employed. It has been well observed * that "brisk and constant markets are to production very much what solar heat and light are to vegetation. A vigorous and sustained demand being given, the expansion of production and the resort to new inventions is as certain as that tropical heat will occasion tropical luxuriance."

The recent discoveries of gold furnish a remarkable illustration of these principles. Among the great

---

* Tooke and Newmarch's *History of Prices*, vol. vi. p. 217.

changes that these discoveries have produced, not the least striking phenomenon is the rapidity of those changes. Under the influence of this potent stimulant to national vitality, infant communities, an unorganized territory of the United States, an outlying dependency of a remote British colony, the desolate and half-forgotten hunting ground of the Hudson's Bay Company, have sprung with unexampled rapidity into political manhood. In those great centres of business which most readily supplied the wants of the new communities, and from them in ever extending circles, a new life seems to have been infused into every branch of industry. Nor does there seem any danger of a glut of gold. In the half century preceding the discoveries in California it is calculated that the quantity of gold in the world was increased by upwards of two hundred millions sterling. In the nine years consequent upon that event, the increase of gold upon that doubled amount was about 30 per cent. Since that period new gold-fields have been discovered, and new methods of extracting the gold have been invented, and the gross yearly produce of the metal is therefore not likely to have decreased. Yet this great increase of quantity has not brought with it any marked or general reduction in its value. There is probably no other commodity which would give such rapid and decided results, and yet experience no check from its own abundance. The reason seems to be that there is no other commodity which, if greatly increased in quantity, would continue to find an immediate and impatient market. If by an increased yield of lead-mines or of copper-mines the stock of these metals were increased in the same proportion as that of gold has been, the gains of those concerned in the mines would in the first instance increase; and subsequently the price of all those commodities in the production of which lead or copper was used would gradually fall. The market for such commodities is limited; and even within its limits is not always brisk. But there is never any delay or difficulty in finding purchasers for gold. It is an object of universal desire, and consequently it finds a market everywhere. Men know that they can sell it again at their pleasure, and are therefore willing at all times to purchase it. Thus it becomes a universal medium of exchange; and so affords a market for all other commodities, and stimulates their production. This increased production reacts upon the utility of the precious metal. More gold is required to carry on the increased amount of interchange. Such at least has hitherto been the case; and the new channels of industry have up to the present time proved deep enough to absorb the additions now annually made to the stock of gold.*

§ 11. The influence of exchange upon invention arises mainly from the extension which the latter agency gives to wants. One of the most powerful influences upon invention is demand; and demand is as it were collected and concentrated by exchange. Individual desires, like individual efforts, are insufficient to induce the extended use of so costly an auxiliary as invention. But the combination of desires, like the combination of efforts, produces marvellous results. This combination is effected and expressed by means of exchange. It is through exchange that the demand is signified, and that its satisfaction can be obtained. Where many persons are willing to purchase each for himself the use or the product of any invention, a sufficient inducement is afforded to competent persons to devote themselves to the supply of such an invention. Accordingly in the earlier periods of industry invention is comparatively slow. There are always men who will pursue knowledge for its own

---

* See Tooke and Newmarch's *History of Prices*, vol. vi. part vii.

sake; but such persons are at first disposed to slight rather than to seek its practical application. In the last century, before the commencement of the present great industrial period, and even for many years after that period had begun, although scientific men showed much mechanical ingenuity, it was almost exclusively applied to the construction of their own philosophical instruments. A few investigations which promised to be of national importance, and had become traditional among several generations of philosophers, were diligently pursued. Such were the inquiries relating to the longitude, to chronometers, and to the lunar theory. But Davy's researches for the safety-lamp were, even at so recent a date, exceptional: and it was thought hardly becoming that Herschel should make a profit by his telescopes, or Wollaston by his platinum.* At the present day an opposite state of things prevails. The commercial element has become in some respects even a nuisance to scientific inquirers. Ignorant and greedy inventors are on the watch for every fresh discovery; and their audacious cupidity has frequently brought discredit upon undeveloped branches of knowledge. But there are more legitimate indications of the stimulus which exchange, when it is facile, quick, and extensive, gives to invention. A period of great commercial activity in England is always a period of inventive activity. Means are contrived for driving engines at higher rates of speed; additional capacity is given to boilers; methods for saving fuel are introduced, and ingenuity is constantly on the stretch to devise means for cheapening production. Many of the large manufactures in Lancashire and Yorkshire employ skilful mechanics at high salaries for the sole purpose of suggesting improvements in their machinery. It was but a short time since that, under the stimulating influence

---

* *Edinburgh Review*, vol. lxxxix. p. 50.

of brisk markets, the progress of mechanical improvements was for a time so rapid that Mr. Babbage estimated the average duration of the machinery before it was superseded by new apparatus at only three years.†

§ 12. I have observed that any extended co-operation implies by its very terms the presence of a considerable number of persons. This is manifestly the case in the combination of labour; and is hardly less manifestly true when the combined labourers are engaged not together but in separate places at different parts of the common work. But an organized system of co-operation requires something more than the number of labourers capable of producing the greatest possible results. The more extended and complete such a system is, the greater will be the product; and the smaller will be that portion of it which is required for the personal consumption of the labourer. In many cases the labourer never uses any portion of the product upon which he works. The operatives in a cotton mill consume but an infinitesimal portion of the products of their factory. A shipwright may never embark in the vessel that he constructs. The labour of these men is therefore useless to them and waste, unless they have the means of converting it or at least its surplus into the appropriate objects of their desires. The extent of co-operation is therefore limited by the extent of the market. If a sufficient market exist, men find that by pursuing distinct occupations they can with the least amount of labour of which the case admits satisfy their wants. If there be no sufficient market, they will so far as they are able supply their wants by their own direct efforts ; because the exclusive pursuit of one occupation would in such circumstances reverse the ordinary industrial conditions, and would

---

† *Edinburgh Review*, vol. lxxxix. p. 81.

give them the maximum of effort with the minimum of result. Thus co-operation depends upon the existing facilities for exchange. Where there are different occupations each of which contains a variety of subdivisions, each of these occupations or subdivisions of an occupation affords a market for the other. Each class produces something different from the other; and each thus produces more than it would produce without such a distribution of work. Thus the power of exchanging stimulates the division of employments. The diversity of employments leads to a large production; and both the large production and the variety of occupations furnish subject matter for further exchange. In such circumstances co-operation can be carried to its utmost extent. Each function, however trivial, may have its appropriate organ. Each organ, however large, is confined to its particular function. But co-operation depends not merely upon the extent of the market but upon its state. The delay that intervenes between the completion of the work and its sale form the characteristic imperfection of co-operative industry. It is a charge which must be deducted from the gross profits of the transaction. Co-operation pre-supposes a sufficiency of capital, but the standard of sufficiency varies in different circumstances. Where the process of exchange is quick and easy, a comparatively small amount of capital will suffice. Where exchange is slow and difficult, a large amount of capital is required. Thus the cost of production, the effort required to obtain the desired object, is in the one case greater than in the other. The existing facilities therefore for exchange, and their influence as well upon the intensity as upon the extension of the demand, materially affect the cost incidental to co-operation; and so determine in any given circumstances the question of its advantage.

Not merely does exchange render possible co-operation, but it also provides for its efficiency when it has been established. In every organism if we desire to secure for each organ its full and unimpeded operation, an exact harmony between the action of the several organs must be observed. When the business is complicated, it is seldom possible to have one man and no more engaged in any one occupation. The proportions of labour required in each department may not be such as without waste to admit of only one man being employed in any one of them. But since the number of men employed increases the aggregate quantity of the product, there must be a market capable of receiving this increased production. The old example of pin-making will once more serve to illustrate this advantage. "If," says Dr. Longfield,* "there was only one man employed in making points, one man would not be enough, and two would be too many, to make heads. The proportion to keep all in constant employment should be four at points and five at heads; and that would require a market in which 180 millions could be annually sold. As there are several other branches of the trade, a still greater number of workmen of each kind would be required to match each other so as to keep all nearly constantly employed. To find the exact number may amuse those fond of arithmetical recreations. We should express in whole numbers prime to each other the proportion of men required in one branch to those in each of the others; and then the least common multiple of all the antecedents will give the number of men who must be employed in that branch so as to prevent any waste."

---

* *Lectures on Political Economy*, p. 103.

CHAPTER XVII.

OF THE INDUSTRIAL ORGANIZATION OF SOCIETY.

§ 1. I have already noticed that in the very constitution of the social unit, the family, the rudiments of society are found. A single family, irrespective of any other and having a sort of completeness in itself, has among its several members various wants and desires; seeks to satisfy those wants and desires by its labour; and strives to increase the efficiency of its labour by its humble capital, and its rude inventions. The co-operation of its various members is shown partly in the union of their forces, and partly in the separation of their occupations which differences of sex or of age necessitate. But although an isolated family can in some sort supply its wants and so maintain its existence, its state is but one remove as it were from extinction. The powers of the semi-solitary man are never fully developed. Where the spontaneous natural product is not of itself and without further change suited to his wants, the profusion of nature is to him rather a hindrance than an advantage. He has not sufficient means to admit of any appreciable accumulation of capital; and he cannot fulfil, even if he possessed the knowledge or the skill, the conditions for the successful control of natural forces. Even when his children become a help and not a burden, their numbers are too few for any extensive co-operation ; and the structure of the family, the partnership of its members and the community of their interests, does not admit of exchange.

These unorganized groups of families are found in the lowest tribes of savages. Even in these groups there is shown some tendency to coalesce. They sometimes act together for some temporary purpose either of war or of the chase : but as soon as the exigency is passed, their connection is dissolved, and they return to their rude independence. It is only in a state of society, where many separate families reside in the same locality, where their interests have become distinct, and where they entertain mutual friendly relations, that any permanent connection is established and industrial development takes place. It is in such circumstances only that all the conditions of an extensive system of exchange can be obtained. It is only by means of such a system of exchange that capital can be kept in constant activity ; or that co-operation can be made to yield all its advantages. It is only when aided by co-operation and exchange that labour becomes sufficiently productive and sufficiently skilful to admit of the rapid accumulation of capital; or to direct the elemental forces, and to fulfil the costly conditions without which the help of these forces cannot be obtained. It is therefore in society and in society only that all the aids to industry become fully efficient, and that the means of satisfying human wants are consequently most abundant and most cheap.

There is a clear distinction between the domestic and the social connection. The one is union : the other is association. Both contain moral and intellectual elements ; but the order of these elements is in each case inverted. "Founded," says M. Comte, * "chiefly on attachment and gratitude, the

---

* *Positive Philosophy*, vol. ii. p. 141.

domestic union satisfies by its mere existence all our sympathetic instincts, quite apart from all idea of active and continuous co-operation towards any end, unless it be that of its own institution. Though more or less co-ordination of different employments must exist, it is so secondary an affair that when, unhappily, it remains the only principle of connection, the domestic union degenerates into mere association, and is even too likely to dissolve altogether. In society, the elementary economy presents an inverse character, the sentiment of co-operation becoming preponderant, and the sympathetic instinct, without losing its steadiness, becoming secondary."

The formation of society is not accidental or capricious. It springs from fixed principles of human nature and is regulated by fixed laws. From the mental and moral nature of man and from his position in relation to external objects there inevitably results a tendency to associate with other men. This tendency admits of being indefinitely strengthened by exercise. The stronger it becomes, the greater are the advantages that it produces; and the greater its advantages, the greater are the inducements to its practice. "To attribute to it," says M. Comte,* "the formation of the social state, as it was the fashion of the last century to do, is a capital error; but when the association has once begun, there is nothing like this principle of co-operation, for giving consistency and character to the combination." The nature therefore of this increased industrial power that society affords, and the influence that in its turn it exerts in consolidating society will next engage our attention.

§ 2. The object and the terms of co-operation depend in the first instance at least upon men's mutual agreements.

---

* Ib. p. 142.

Such an agreement, whatever its nature or its consideration may be, implies communication between the parties and proposals made by the one and accepted by the other. But the agreement which is thus expressly made between individuals, exchange in effect extends by a tacit understanding through every part of society and to great classes of men. This result arises spontaneously from the circumstances of society, and from the opportunities for exchange which a state of society implies. The process by which the result is attained may be readily traced. A man acquires, whether from natural talents or from some accidental circumstance, the power of rendering some service with unusual skill. At first he follows this occupation in conjunction with some other pursuit as his circumstances require. By degrees his powers become known, and many persons desire his assistance. His practice soon brings increased skill, and his new skill tends further to increase his practice. At length he finds the number of his clients so much increased that it is profitable to devote himself exclusively to the supply of their requirements. The public on their side feel that in this matter they can be better and more economically served by employing the practitioner than by attempting to render the service for themselves; and are consequently willing to pay according to the nature of the service. The successful practitioner gradually becomes rich. Other persons are encouraged by his example, and imitate it. Successors or competitors soon appear, and thus a regular business is established. Guided by experience, men speculate upon the probable wants of their neighbors, and prepare themselves to satisfy those wants. They know that these neighbors will desire to have certain services rendered to them; and that in consideration of the trouble thus saved they will cheerfully pay a reasonable reward. The farther this arrangement is carried, the stronger the inducements to its extension become. Each occupation becomes distinct from, and at the same time more or less dependent upon, every other occupation. The skill resulting from constant practice continually places the men of special beyond the men of universal art: and each art thus specialized forms a market for the labour of those who pursue other occupations. Each class exchanges with every other; and thus the whole society spontaneously and without any design on the part of its members assumes a co-operative character. Every member of society is dependent on the aid of others in all his acts and for all his enjoyments. "All society," it has been said,* "is in fact one closely-woven web of mutual dependence in which every individual fibre gains in strength and utility from its entwinement with the rest. But while all the members of society co-operate for a common purpose, the increase of the general welfare, each individual is still strictly occupied in pursuing what he considers his own private and exclusive interest in whatever way he likes best."

§ 3. Of the gain thus obtained there is no doubt. The acquisition of the benefits that result from co-operation is determined by the fact of co-operation, and not by the mode of its origin. They are the same whether the action be concerted or spontaneous. But these benefits, great as they are in the case of a small body of workers, are indefinitely multiplied when that body expands into a nation. The construction of the simplest agricultural implement, a spade or a pair of shears, implies a great variety of arts. Each of these arts, and each again of these latter arts, involves numerous other branches of industry, which themselves too expand as they arise, until the production of such a simple tool seems to require for its antecedent the whole course of civilization.

---

* Scrope's *Political Economy*, p. 76.

Thus Adam Smith's remark is strictly true that "without the assistance and co-operation of many thousands the very meanest person in a civilized country could not be provided, even according to what we very falsely imagine the easy and simple manner in which he is commonly accommodated."

The great advantage which society brings to the producer is an extended market. A number of moderate payments soon surpasses the bounty of the most wealthy and the most munificent patron. Such patrons are necessarily few; but the supplies that the public can afford are abundant, and are constantly increasing. No individual however wealthy could maintain a staff of literary men, of artists, of engineers, and of manufacturers. If he were able to do so, he could not find employment for them. But when the public is their master, all these classes find full work and ample payment. What is perhaps still better, they are independent. They are not compelled to shape their services so as to humour the caprice or to flatter the vanity of an individual. What a difference, not only pecuniary but moral, exists between the relations of a successful modern author with his publishers, and the relations of Dryden or of Savage with the persons of quality for whom they wrote dedications or epitaphs! What the booksellers have done for the authors, the print-sellers have done for the artists. In like manner the young Levite of two centuries ago has been developed into the popular preacher of the present day. The work for which an engineer now receives ten guineas a-day was cheerfully done by Brindley for a daily wage of two shillings; and Sir Christopher Wren were he now alive would not be required daily to be drawn up in a basket to the dome of St. Paul's for no higher salary than £200 a-year.*

On the other hand the benefit to the community is not

---

* See Smiles's *Lives of the Engineers*, vol. i. p. 324.

less remarkable. The sum which in the aggregate forms a splendid remuneration, falls lightly upon each of its contributors. When this cheap method of obtaining services becomes known, the number of customers, according to a principle already illustrated, increases. By the usual reaction the reduced price attracts greater numbers; and the increased numbers render possible a reduction of price. Thus civilization tends as it has been well observed,* not to make bad things for nothing but to make good things cheap. The meanest labourer in a civilized community can thus satisfy his primary wants, can obtain food and clothing, shelter and warmth, far more efficiently and constantly than "many an African king, the absolute master of the lives and liberties of ten thousand naked savages." But society does for the poor man, that is for the mass of the population, more than this. It enables him to share in the enjoyments of the rich, and to obtain the same services that are at their command. The humblest charwoman can have her tea, the poorest hodman can have his tobacco, in the quantity, of the kind, and at the time that they desire, with as much certainty as a Royal Duke. The same telegraph bears the message of the sovereign and the tradesman. The same train conveys the peer and the peasant. The surgical assistance of a Paget, and the forensic abilities of a Palmer are within the reach of men of very moderate means. Few persons in any of the great European cities have been too poor to enjoy the dramatic talents of Fechter or of Kean, or the vocal powers of Grisi or of Lind.

When society is imperfectly developed, even though there be great individual wealth, there is a marked absence among every class of the comforts that are due to civilization. Men will not incur the entire expense

---

* Chenevix on *National Character*, vol. ii. p. 70.

of advantages in which from their very nature others who have not paid for them must share. They will rather, notwithstanding the discomfort, forego the undertaking. However rich a man may be, and however he may be inconvenienced by their absence, he will not at his own cost build public bridges or make public roads. When there are but a few wealthy and scattered residents in a district, there may perhaps be found in that district good houses and handsome furniture, servants and horses, beautiful pictures and costly wines: but there will be no roads, no bridges, no schools, no churches, no libraries, no local newspapers, no theatres, or gardens, or other places of public amusement or resort. There will not be "that subtle force and discipline which comes of the myriad relations with and duties to a well constituted community which every member of it is daily exercising, and which is the natural unseen compensation and complement of its more obvious constraints and inconveniences."* Mr. Olmsted, from his personal experiences in the slave states, forcibly illustrates the results, in perhaps their worst form, of such a state of society. He frequently dwells upon the entire absence of comfort in most Southern houses, and the small portion of the conveniences of life that in these countries riches bring to their possessors. Men there often become rich in a few years, and own many negroes, and raise many bales of cotton. "But in what else," he asks,† "besides negroes were these rich men better off than when they called themselves poor? Their real comfort, unless in the sense of security against extreme want, or immunity from the necessity of personal labour to sustain life, could scarcely have been increased in the least. There was at any rate the same bacon and corn, the same slough of a waggon channel

---

* Olmsted's *Journeys in the Cotton Kingdom*, vol. i. p. 21.
† Ib. vol. ii. p. 302.

through the forest, the same bare walls in their dwellings, the same absence of taste and art and literature, the same distance from schools and churches and educated advisers, and, on account of the distance of tolerable mechanics and the difficulty of moving without destruction through such a rough country anything elaborate or finely finished, the same make-shift furniture."

§ 4. There are few phenomena within the range of human experience more remarkable than the readiness and the certainty with which when a demand is strongly felt new agencies for its supply spring into existence. In every kind of undertaking, from the most gigantic to the most humble, from the most complex to the most simple, there is the same anticipation of human wants and the same eagerness to supply them. In the ordinary branches of industry examples of this tendency are so familiar that they fail to awaken the attention they deserve. It is among new communities, or in a social state wholly dissimilar to our own, that we can perceive the readiness with which these social agencies spontaneously adapt themselves to every variety of circumstance. It matters not how novel or how repulsive the character of the service may be, if those who require it be sufficiently numerous, that service will surely be rendered. Thus in parts of India there is a class of professional thief-takers. These men trace thieves by their footsteps, and are said to equal in their calling the unerring sagacity of the bloodhound. Guzerat is celebrated for people of this class. Notwithstanding the faintness of the trace that a naked foot leaves upon a dry soil, one of these men "will perceive all its peculiarities so as to recognize it in all circumstances, and will pursue a robber by these vestiges for a distance that seems incredible."* In Alabama and other Southern

---

* Elphinstone's *History of India*, p. 192.

States, there are men who make the pursuit of runaway slaves their special business.* Dogs, generally of the bloodhound kind, are trained to hunt negroes; and the owner advertises his abode and business in the usual way. The remuneration for the service varies from ten to two hundred dollars, according to the time occupied in the chase. In the same country there are regular flogging establishments for slaves. When a master desires to have his slave punished, he sends him to one of these houses with an order stating the amount of punishment required; and the man is flogged accordingly at current rates. The same kind of accommodation appears in similar circumstances to have been provided for the convenience of masters by the enterprising speculators of Ancient Rome.

We turn from such examples to the legitimate employments which the new industries incident to new countries have produced. A striking instance of this class is found in the settlement of waste lands. Without any precedent to guide them, without any direction or assistance from the state, but merely in pursuance of their own feelings and views of convenience, the American settlers have made the preliminary clearing of the wilderness a distinct occupation. So complete is the separation of function, and so great is the skill that the persons who follow this business have acquired, that no advantage of capital will enable an ordinary settler to compete with them. "It is impossible," says Mr. Charles Buller,† "to conceive a more striking contrast than is furnished by the present state of settlements thus formed by persons who had no property when they entered the bush but an axe and a camp kettle, and that of

---

* Olmsted's *Journeys in the Cotton Kingdom*, vol. ii. p. 120.

† *Report on Public Lands*, p. 32, Appendix B. to Lord Durham's *Report on Canada*.

settlements formed by British emigrants possessed of considerable capital. The Americans have almost uniformly prospered. The European emigrants have always been slow in their progress, and have not unfrequently been ruined. . . . . Indeed there appears to be in this, as in almost every other pursuit, a natural division of employments; and this is practically understood in all parts of the United States. One class of persons attach themselves almost entirely to the occupation of breaking up new land. They go into the wilderness, select a favourable location, erect a small hut and commence the task of clearing. In a few years the progress of settlement brings other settlers into the neighbourhood; and they then sell their improvements, and again move off several miles in advance of the tide of population, repeating the same process as often as they are overtaken by it. By their labours the difficulties of a first settlement are to a great extent obviated. Those who succeed them are spared the worst and most disheartening part of the toils of a settler; and the work of settlement proceeds more rapidly and prosperously than would be the case if those who eventually occupy the land had been also the persons by whom it had been first reclaimed."

A similar case, less complete perhaps than in America, occurs in Victoria. There is in this country a class of men who devote themselves mainly to the search for gold fields. These prospectors, as they are called, are not regular miners; that is, they do not derive their livelihood, to any great extent at least, from the extraction of gold. Not unfrequently indeed they are miners who have been by some casualty prevented from following their occupation, but who have some skill in detecting those appearances which indicate to an experienced eye the probable presence of gold. They therefore do not remain at any one gold field, but are continually on the move in search of some new discovery. When they have discovered a promising claim, they sell their interest to those who follow them, and who desire to search for gold in the ordinary manner. When they find not merely a superior claim but a new gold field, they are entitled under the mining laws to receive an extended claim : and this claim they in like manner sell, and then recommence their search. Thus the actual miner is saved the loss of time and trouble and the expense of searching for a favourable locality for his operations. He quietly and without delay works at the claim that has been found and tested for him ; and the prospector goes on his way to render a similar service to some new customer. In the more remote parts of the Australian continent the same practice prevails, and in the older colonies during the early days of their pastoral settlement formerly prevailed, in respect to runs. Men applied themselves to the discovery of new country suitable for pastoral purposes : when they had found and secured a run, they sold their interest to some intending settler ; and then, like the backwoodsman or the prospector, they resumed their former search.

§ 5. Not less striking than the promptitude with which the supply of services responds to the demand is the precision with which their succession is regulated. In any artificial arrangement of a similar kind questions would necessarily arise as to the priority of the services that should be rendered. It would not be easy to find any authority competent to decide such questions; and even if such an authority were found, his decision would seldom be generally accepted. It would be necessary to settle beforehand what each class should do or forbear to do ; and causes of discontent would arise as well from interference as from neglect. In the order of nature all these questions are conclusively settled by an unfailing rule. All services are rendered according to the order of their social importance. Those which society at the time most needs will be rendered first. Those which are not immediately essential to its welfare will come in due time. The process is as simple as it is efficient. "Each man does that which pays him best: that which pays him best is that for which other people will give most: that for which they will give most is that which in the circumstances they most desire. Hence the succession must be throughout from the more important to the less important. A requirement which at any period still remains unfulfilled must be one for the fulfilment of which men will not pay so much as to make it worth any one's while to fulfil it; must be a less requirement than all the others for the fulfilment of which they will pay more; and must wait until other more needful things are done."*

Not merely do these social arrangements appear with unfailing certainty when they are wanted, and in the order which the intensity of our desires indicates; but they last so long as they are required and no longer. The same principle which prevents the growth of any superfluous or premature occupation secures the proper duration of those which last, and their speedy and effectual removal when their utility is gone. New trades and new companies daily start into existence: if they supply some existing public want, they prosper: if they do not supply such a want, they fail. In the natural organism, the organs, while they discharge their functions, attract sufficient nutriment, and are healthy: if their functions be limited or cease, the nutriment is no longer supplied, and they dwindle away or die. In the social organism a similar process takes place. If his function be really useful, the single worker attracts other workmen to him, and

---

* Spencer's *Essays*, p. 340.

becomes the nucleus of a great establishment; and this establishment by degrees expands into a complex system of separate trades.

In the extension also of these arrangements the same capacity of adaptation is observed. Under the same stimulus that excited its growth, industry will branch out in whatever direction and to whatever extent its circumstances require. On the other hand, if the worker fail to secure or to retain public support, he seeks for some want which is more keenly felt than that which he formerly sought to supply; and is speedily taken up and absorbed by some of the many employments to which society gives rise. These changes too are for the most part gradual and consequently sure. "There is no part of any organism whatever but begins in some very simple form with some insignificant function, and passes to its final stage through successive phases of complexity. Every heart is at first a mere pulsatile sac; every brain begins as a slight enlargement of the spinal chord. This law equally extends to the social organism. An instrumentality that is to work well must not be designed, and suddenly put together by legislators, but must grow gradually from a germ : each successive addition must be tried and proved good by experience before another addition is made, and by this tentative process only can an efficient instrumentality be produced."* Except in rare cases of accident or error, ruin or success do not come all at once. Hence, when failure is probable, the worker has practically notice that his services are not required; and has sufficient opportunity to select some more promising occupation. The successful man on the other hand is not only stimulated to fresh efforts, but by his previous training is fitted for them. By the combined influence of these forces, failure and success, each step in the social advance is separately

---

* Spencer's *Essays*, p. 332.

tested, so that no weak or premature portion can remain ; and when new work requires to be done, the existing agencies fit themselves for their new duties and readily undertake them.

§ 6. When in any large community industry has been thus spontaneously organized, and when the facilities for exchange between the various parts of the community have increased, a new phenomenon presents itself. The same separation of functions which takes place between different occupations, takes place also between different localities. The various branches of industry exhibit a strong tendency to fix themselves in, and confine themselves to, particular districts. Each district thus acquires a distinctive character, and at the same time becomes dependent upon the other districts with which it deals. In England this localization of industry is peculiarly marked. The manufacturers of cotton, of wool, of stuff, and of silk, the workers in hosiery and in lace, in coal and in iron, in cutlery and in earthenware, in straw-plait and in needles, and in a thousand other occupations have each their separate locality. In no place perhaps is it so conspicuous as in London. "Lawyers," it is said,* "live in Kensington district in greater relative numbers than in any other district, but the law-clerks, except those who live round Chancery Lane and the Inns of Court, are found in greater relative numbers at Islington. The authors, editors, artists, and architects, are found in small numbers South of the Thames or in the Eastern half of the Metropolis. Marylebone, St. Pancras, and Kensington are their chief districts. Domestic servants are found in greatest relative force in the districts of St. George's, Hanover Square, St. James', Westminster, Marylebone, and Kensington, indeed

---

* *Companion to the Almanac,* 1855, p. 80.

X

overwhelmingly so. The tailors are strong in St. James', Marylebone, and St. Pancras, but relatively more so in the Whitechapel and neighbouring districts where much of the slop work is done. The chief districts for shoemakers are St. Pancras, and Marylebone in the North, Lambeth and Newington in the South, Whitechapel and Bethnal Green in the East. The gardeners have Kensington and Wandsworth as their chief localities. Beyond all other districts the City is the locality for publishers and booksellers, for it contains the regions of Paternoster Row, and the numberless courts around Fleet Street. Musical instrument makers congregate in decided preponderance in St. Pancras. Watchmakers appear in surprising force in Clerkenwell and St. Luke's. Coachmakers in St. Pancras and Marylebone: shipbuilders in Stepney and Poplar: dyers and calenderers in Shoreditch and Bethnal Green: leather-workers in Bermondsey: sugar-refiners nearly all in Stepney, Whitechapel, and St. George's in the East: cabinet and furniture-makers in Pancras, and especially Shoreditch: coopers in the districts nearest the various docks: rope and sailmakers in Stepney and Bethnal Green: workers in gold, silver, and precious stones in Clerkenwell."

In Holland, during the days of its commercial greatness, various branches of commerce selected as it were each some favourite town. Middelburg was the great seat of the wine trade. Flessing almost monopolized the East India trade. Shipbuilding was the chief business of Saardam. Sluys was devoted to the herring fishery.* In the Tartar villages and in most Oriental towns, just as in London itself, the artizans live together according to their trades. The same tendency is curiously illustrated in Russia. In the remarkable communal system which

---

* Blanqui's *Histoire d'Économie Politique*, tome ii. p. 32.

forms the basis of society in that empire, the various trades have formed themselves into separate communes. All the inhabitants of one village for example are smiths, of another carpenters, of a third curriers. In one village nothing but boots and shoes is produced; in another earthenware; another is exclusively devoted to training birds and the bird trade; while in others the whole population consists of beggars.* The Russians are accustomed to live together in large families; and so an industrial organization has been established in each family, which continues and is extended in that union of families which forms a commune.†

§ 7. The cause which determines this localization of industry, although it may often be difficult from the want of proper local knowledge to detect its presence, is some economy in the cost of production. From some circumstance, sometimes natural, sometimes arising from the character of the inhabitants, sometimes purely accidental, one district is found to possess unusual advantages for some particular occupation. Thus the facility of obtaining the raw material has determined the position of many leading branches of industry. Cotton-mills are found in the vicinity of seaports: iron-works where iron is interstratified with coal: the worsted trade has flourished in those districts where the long-woolled breed of sheep has been most successfully reared: cloth-weaving, as long as it was handicraft, remained in the countries where the fleeces were soft and short in the staple: and the potteries were established in the vicinity of the most ductile clays. But neither the mere presence nor the mere absence of the raw material is

---

* Haxthausen's *Russian Empire*, vol. i. p. 190, 141.
† Ib., p. 54, 154.

without reference to other considerations sufficient to determine the locality of most industries. In many manufactures that which appears to be the principal material is in reality only one out of several different requisites; and the facility or the difficulty of obtaining these requisites or some of them may more than compensate for a deficiency or an advantage in the principal raw material. The manufacturers of porcelain and earthenware at Glasgow are obliged to import their clay, their flints, and Cornish stones from Dorset, Devon, and Cornwall. The burden of transporting their raw material from such a distance would seem to preclude all hope of success in such an undertaking. But there are many other articles besides clays and flints used in this manufacture; and Glasgow has peculiar facilities for obtaining many of those articles, while her advantages with respect to foreign commerce surpass those of her competitors.*

Again, when power is applied to production, the choice of locality depends upon the result of the inquiry where coal or water can be had at the cheapest rate. The skill, or at least the aptitude, of the population, and in some cases their willingness to use the machinery employed, are also important circumstances. When the seat of the industry is once established, even though its origin may have been merely accidental, various further subdivisions of employment arise among the inhabitants ; and thus as the organization is rendered more complete than elsewhere, the local superiority becomes more marked. Where machinery is used, the assistance of engineers is required ; and where many machines are engaged, such services find a profitable market. In such circumstances a manufacturer can at a small cost have his engines repaired and kept in order ; if on

---

* *Journal of Statistical Society*, vol. xx. p. 182.

the contrary he were to settle in another locality, he would be obliged either to send his machinery to a distance for every trifling repair, or to burden his business with an engineering staff for which he had not full employment.

Another advantage from the presence of skilled mechanicians is the continuous though hardly perceptible progress of mechanical improvements. "There is hardly a factory," it is said, "in which the machinery does not receive some change or alteration every month, small indeed in amount but great in importance when reference is made to the mass of production over which its influence ranges. A cotton-mill, in the machinery of which no changes had been made for twenty years, would be left so far behind in the march of improvement that it could hardly be worked with profit. Now where several factories of the same kind are aggregated, there is a constant demand and an adequate remuneration for invention; but this is not the case where a factory isolated and unconnected is for the first time established. It is to this cause mainly that the cotton-spinners of Lower Normandy attribute their inferiority to their English rivals, for though the free export of machinery is permitted, yet after a few months some improvement, easily effected by some slight modifications in a place where mechanicians abound, gives the Manchester manufacturers a start, and the French cannot get up to them without importing an entire set of new machinery."*

Another cause which sometimes affects the locality of a manufacture is the requirements of the staple business of the place. As in a factory the business is frequently extended with the growth of the establishment until it includes a variety of distinct but subordinate occupations, so

---

* W. C. Taylor LL.D.: *Transactions of Dublin Statistical Society,* vol. i. p. 6.

in a locality a new business is sometimes called into existence to supplement some more important trade. Such was the origin of the Glasgow potteries, to which I have already referred. When the foreign trade of the Clyde grew up, it was found that the locality could supply abundance of heavy freight in the form of iron and of coal, and also of fine goods; but there was no bulky freight to fill the room that was still unoccupied. The rival port of Liverpool had long enjoyed from the potteries of Staffordshire this advantage; and the merchants of Glasgow with this example before them perceived their interest in developing a trade that gave full employment to their shipping.*

§ 8. "Were all nations to follow the liberal system of free exportation and free importation, the different states into which a great continent was divided would so far resemble the different provinces of a great empire. As among the different provinces of a great empire the freedom of the inland trade appears both from reason and experience not only the best palliative of a dearth, but the most effectual preventive of a famine: so would the freedom of the exportation and importation trade be among the different states into which a great continent was divided. The larger the continent, the easier the communication through all the different parts of it both by land and water, the less would any one particular part of it ever be exposed to either of these calamities, the scarcity of any one country being more likely to be relieved by the plenty of some other. But very few countries have entirely adopted this liberal system." Notwithstanding the force and the frequency of the impediments to which Adam Smith thus alludes, the tendency is not on that account the less certain; and its extent is by no

---

* *Journal of Statistical Society*, vol. xx. p. 134.

means limited to the case of food. Its existence indeed can hardly be seriously denied. England obviously produces all those commodities which owe their cheapnesss to her abundant mines of coal and of iron. The vine and the silkworm supply the industries which predominate in Southern Europe. In America the Southern States produce cotton; the Western States raise agricultural produce; while the principal seats of manufacturing industry are found in the North and in the East. The West Indian Islands confine their labours to the production of sugar. The Chinese are still unrivalled for tea. Among the Australian group of colonies the staple industry is in Victoria gold, in New South Wales and Queensland wool, in South Australia breadstuffs. So strong is this tendency towards a territorial division of employments that human efforts, although they may sometimes check it or even change its direction, are never able altogether to conquer it.

§ 9. The motives which determine the organization of industry as between different communities are the same as those which determine its localization in different parts of the same community. Both phenomena are merely cases of exchange between individuals; and, except so far as the difference of laws or national policy may disturb that exchange, the nature of the transaction is not affected by the common or the different nationality of the parties. If the parties be free from any interference, the same economy in the cost of production which induces the workers in clay to settle in Staffordshire, and the workers in cotton fabrics to congregate in Manchester, will influence the Chinese to confine themselves to the cultivation of the tea plant, and the West Indians to the cultivation of the sugar cane. The most conspicuous instance of this economy is found of course in the physical differences between various countries. These differences are more strongly marked than even those between individuals, and have been at all times noticed. In a well-known passage, Virgil, when he warns his husbandman to make himself acquainted with the winds and the changeful humours of the sky, and the mode of culture and the character peculiar to each place, and to ascertain what each district will produce and what it refuses to bear, appeals to the laws and everlasting covenants which nature has from the very first set for different countries. He tells how Tmolus yields the fragrant saffron, India its ivory, the gentle Sabæans their frankincense; how the naked Chalybes send their iron, Pontus the strong-smelling Castor, and Epirus the mares that win the Elian palm. With still greater emphasis is the same principle asserted, and the incidental advantages of commerce indicated in a document not indeed of such poetic beauty as the Georgics, but to us of a great and enduring interest. In the circular letters of commendation given by the ministers of Edward VI. to our early navigators, Sir Hugh Willoughby and Captain Richard Chancellour, and addressed to all kings, princes, and persons in authority, the following passage, remarkable when we remember the time at which it was written, occurs :—" And if it be right and equity to shew such humanity to all men, doubtless the same ought chiefly to be showed to merchants who wandering about the world search both the land and the sea to carry such good and profitable things as are found in their countries to remote regions and kingdoms; and again to bring from the same such things as they find there commodious for their own countries: both as well that the people to whom they go may not be destitute of such commodities as their countries bring not forth to them, as that also they may be partakers of such things whereof they abound. For the God of Heaven and Earth greatly providing for mankind would not that all things should be found in one region, to the end that one should have need of another; that by this means friendship might be established among all men, and everyone seek to gratify all."*

But natural advantages are not with nature any more than with individuals the sole or even the chief cause that determines the direction of industry. That direction is very frequently determined by the proximity of markets, or by the skill or character of the inhabitants, or by some of those other circumstances which I have already noticed. In many seas there are great fishing banks, in many rivers abundant water power, in many lands rich minerals or noble forests; but the economic conditions for their profitable use have not yet been fulfilled. Nor is the due compliance with these conditions the sole consideration. Even where the natural advantages determine the character of either individual or national industry, it is not the absolute advantages enjoyed by that individual or that nation, but its relative advantages that we are to estimate. A man who earns £10 a-day as a physician, even though he may himself be a most skilful compounder, will not mix his own drugs, if he can get another person for a mere fraction of that sum to render him that service. In like manner, if the inhabitants of any country can raise one product to the value of ten millions annually, they will not in preference raise other products to only half that amount, notwithstanding that their production would even thus exceed that of their neighbor. A good example of this principle is given by an American writer,† "The inhabitants of Barbadoes favoured by their tropical climate and fertile soil can raise provisions cheaper than we can in the United States. And yet Barbadoes buys nearly all her

---

† Quoted in Mr. McCulloch's Introductory Discourse to his Edition of the *Wealth of Nations*, p. xxv.
† Bowen's *Political Economy*, p. 460.

provisions from this country. Why is this so? Because, though Barbadoes has the advantage over us in the ability to raise provisions cheaply, she has still a greater advantage over us in her power to produce sugar and molasses. If she has an advantage of one quarter in raising provisions, she has an advantage of one half in regard to products exclusively tropical: and it is better for her to employ all her labour and capital in that branch of production in which her advantage is greatest. She can thus by trading with us obtain our breadstuffs and meat at a smaller expense of labour and capital than they cost ourselves. If for instance a barrel of flour cost ten day's labour in the United States, and only eight day's labour in Barbadoes, the people of Barbadoes can still profitably buy the flour from this country if they can pay for it with sugar which cost them only six days' labour, and the people of this country can profitably sell them the flour or buy from them the sugar, provided that the sugar if raised in the United States would cost eleven days' labour. This is a striking example to show the benefit of foreign trade to both the countries which are parties to it. The United States receive sugar which would have cost them eleven days' labour by paying for it with flour which costs them but ten days'. Barbadoes receives flour which would have cost her eight days' labour by paying for it with sugar which costs her but six days'. If Barbadoes produced both commodities with greater facility, but greater in precisely the same degree, there would be no motive for interchange."

CHAPTER XVIII.

OF THE ADJUSTMENT IN SOCIETY OF THE TERMS OF CO-OPERATION.

§ 1. In the simpler forms of co-operation the adjustment of terms is comparatively easy. The parties readily discuss and settle the terms of their partnership. Their number is small, and the complexities are few: there is consequently little room for dispute; and the differences, where they do exist, are palpable. In such circumstances, men quickly make such agreements as the nature of each case requires. But the case is altered when we look at that great system of tacit co-operation which has grown up with society. There is no way of bringing together different classes of workers. There can be no mutual discussion of terms; no arrangement as to the time and the mode of payment of their respective shares, or as to the system of management which their common interest requires. Men who live often under separate jurisdictions have no common means for settling their disputes; and even when the same tribunals are open to both parties, such quasi-contracts are too vague to admit of interpretation and enforcement. Since then the rules by which the terms are settled in ordinary partnerships are manifestly inapplicable to these great social agencies, we must inquire in what manner this essential part of co-operation is in cases of tacit co-operation carried into effect.

The mode in which the difficulty is met is by the combined action of capital and of exchange. The person who advances the capital discounts as it were the share of those who contribute only labour. He purchases from them for ready money their interests in the undertaking. This purchase is of course an exchange in which for a stipulated amount all the other co-operators transfer their rights to one person. He thus acquires the property in the undertaking, with all its chances of gain and all its chances of loss. He by some act or some forbearance effects some alteration in the subject either as to form or time or place; and so increases the utility of the service rendered. Subsequently he transfers the whole of his interest to another capitalist, who in his turn transfers the property to a third; and thus it passes through successive hands, each proprietor receiving from his immediate predecessor his own share and the shares of all the others which he advanced. Each alteration in the product that is effected in these various stages implies some share, however minute, in some previous co-operation; each of which shares in turn involves other such shares; and so there is constituted a chain of wages and profits that extends to the earliest dawn of civilization. The price therefore that the ultimate purchaser pays represents the remuneration for the total amount of services rendered at different times by different persons in the preparation of that service thus rendered to him. It is by this combination of capital and exchange that that unconscious co-operation is effected, by which a pound of cotton wool, worth at the most two shillings, is transmuted, without difficulty and to the satisfaction of all parties, into a pound of lace worth one hundred guineas. "No means," says Mr.

Senior,* "except the separation of the functions of the capitalist from those of the labourer, and the constant advance of capital from one capitalist to another, could enable so many thousand producers to direct their efforts to one object, to continue them for so long a period, and to adjust the reward for their respective sacrifices."

§ 2. But the number of partners and the consequent difficulties are not the only causes which embarrass the arrangement of the terms in co-operation. The assistance which the co-operators respectively render may vary both in degree and in kind. Their contributions to the common stock may not all be either equal in amount or uniform in kind. From what I have already said the various classes of difference may be readily perceived. When one man seeks the aid of another, there are several ways in which that other may aid him. The co-operator may supply either the natural agents that he has appropriated; or the inventions at his disposal; or his capital; or his personal exertions. When the co-operation is extensive, all of these aids or any of them may be contributed in equal or in varying amounts by all or by some of the co-operators. Sometimes for example a company is formed where each member contributes a portion of each of these aids. Sometimes an employer possesses natural agents and capital and invention, and requires merely personal attendance. Sometimes again a worker starts on rented land, and with borrowed capital and inventions; and sometimes he himself possesses one or more of these auxiliaries, and is indebted for the remainder to some accommodating friend. But whatever may be the facts in any particular case, it is important to ascertain how the possession of any of these auxiliaries affects the position

---

* *Political Economy*, p. 81.

in the partnership of the owner. The share of each contributor is naturally determined by the amount of his contribution. This rule, although it may be sufficient for homogeneous shares, fails to meet the case where the nature of the contributions differs. By what principle are the relative shares of the capitalist and the labourer fixed? What portion, if any, of the ultimate price is due to nature? Who receives this amount and what is his warrant? To these questions some answer must be attempted. I proceed therefore to consider the laws which govern the remuneration given for natural agents, inventions, capital, and labour, respectively.

§ 3. For the present purpose no difference needs be made between natural agents and inventions. Both these terms imply the presence of physical agencies working with the industry of man. The particular manner of the combined action is not now material. But such action, whatever form it assumes, neither requires nor admits of price. Labour is the purchase-money of all things. Some human effort, whether it be great or small, whether it be made by the recipient of the gratification or by some other person in his behalf, must be made before the satisfaction of any human desire. We are surrounded by objects capable of gratifying our desires; but we must stretch forth our hands to gather them. The payment of this purchase money is essential, but it is sufficient. Man must conquer nature, and conquer her by obedience; but the conquest once made is complete. As soon as the duly directed effort has been made, nature yields all her energies, whether they are the recipients of human impress or the substitutes for human force, without further condition or reservation, without money and without price.

Of the reality of this surpassing gift, of the fact that

in human industry the co-operation of nature is always gratuitous, there is abundant proof. Whatever may be the difference in all other respects of the various natural objects which command a price, whether they be of primary necessity to man or merely idle toys, whether they come from earth or ocean or air, whether they be living creatures or mere inanimate forms, they all alike agree in one particular, that they have been procured, or rather that similar objects if they be desired must be procured, by human labour. But there is still more cogent proof that it is labour and labour only for which we pay in the purchase of any commodity. The same commodity which in its natural state is without price, becomes saleable when altered in its position or its form by the hand of man. Ice bears no price on Wenham Lake; but it finds a ready market in Calcutta or in Melbourne. The same ice which during summer is in such demand, finds no market during winter. Oxygen is universally diffused throughout the world; but if we desire it in its pure form, we must pay the ordinary wages of chemical skill. There are thus two cases similar in all respects save one, and with a dissimilar result. The novel consequent must therefore be attributed to the novel antecedent; and we are entitled to say that the cause of the price is the necessity of labour. If further proof be desired, price varies with and according to the effort necessary to attain the object. Water at the river side costs merely the trouble of appropriation. The only effort there requisite is that of stooping and drinking it. When the same water is conveyed to an adjacent town, it costs a certain sum. If it be brought beyond the town, its price increases with the distance. But if the town be supplied by water works, in which the agency of nature is substituted for the agency of man, the price will be greatly reduced. As the labour increases, the price increases; as the labour diminishes, the price diminishes; but in all cases, and with every variation, it is the same water, and the same utility to man.

§ 4. There are several circumstances which obscure this important truth. Our attention may be directed in some cases to the result of the labour, in others to the act of labour itself. A man enters a tailor's shop, and buys a ready-made coat, and leaves some cloth that he has brought with him to be made into another coat. In the former case, he says that he pays for the garment; in the latter, that he pays for the making of the garment. But in both cases, if we omit for the moment the preparation of the cloth, he really pays for the same thing, the service rendered by the tradesman in the preparation of the article. This impression of a change in the character of the service is greatly strengthened by a change in its name. A shoemaker, as Mr. Senior has observed, changes leather into shoes. A shoeblack changes dirty shoes into clean shoes. But in common language we say that we buy the shoes, and employ the shoeblack. In the former case, we look to the result of the labour embodied in the commodity; in the latter, the attention is directed to the act of labour exerted upon the external object. In both cases, there is the labour and there is the natural agent; but that which was primary in the one case becomes secondary in the other. We are also greatly influenced on this subject by the diversity of employments. When a man foresees that certain services which he can render are likely to be required, he in anticipation of the demand prepares himself in advance so as to be in a position to render these services. A merchant foresees a demand for tea, or for wine, or for hardware; and lays in a supply of these commodities. The purchaser pays apparently for the commodity, but really for the service rendered to him by the merchant and by the other producers, all which services are embodied in and represented by the commodity. Such a commodity, in the words of M. Bastiat, is merely "service prevu."

M. Bastiat has traced the various steps by which the service is gradually merged in the commodity. Two neighbors require water; with a view to that economy of labour which is at the root of all exchange they agree that one shall bring the water requisite for both, and that the other shall teach the child of the former. The service of tuition is the payment for the service of carrying water; and the one service exchanges for, or is worth, the other. Subsequently the teacher wishes to discontinue his occupation, but at the same time to receive the service of the water carrier. He proposes a new consideration, and offers money. Water-carrying is then worth so much. In a short time the water carrier no longer waits for a request to render his services; but as he is able to calculate upon a certain demand for them, he prepares to supply water to those that require it. He becomes a water seller. Then it is said that water is worth so much. The barter of services, the sale of a service, the sale of a commodity, such are the several stages of the transaction. "And yet," asks M. Bastiat,* "has the water really changed its nature? Is the value, which hitherto consisted in the service, materialized, by being incorporated in the water, and adding to it a new chemical element? Has a trifling modification in the form of the arrangements between my neighbor and myself the power of displacing the principle of value, and of changing its nature? I am not such a purist as to object to any one saying that water is worth sixpence, just as one says that the sun sets. But it must be understood that these expressions are merely metaphorical; that metaphors do not

---

* *Harmonies*, p. 112.

affect the reality of facts, and that, scientifically, value no more resides in the water than the sun sets in the sea."

§ 5. In the great majority of cases there has been little difference of opinion upon this gratuitous character of nature's co-operation. But as the epithet gratuitous implies the absence of price, and as price is a form of exchange, and as exchange can only take place where attainment is difficult, it was soon observed that this gratuitous character must be confined to those cases where the quantity of the desired object is practically unlimited. Since price thus made its appearance only where the supply of natural agents was limited, it was inferred that in such cases the price was paid for the use of the appropriated natural agents; that the price so paid constituted rent; and that, as the natural agent most susceptible of appropriation was land, " the great monopoly of land " forms an eternal and immense exception to our rule. I have already endeavoured to show that there is no real difference between land and any other natural agent: yet so inveterate are our associations and so great the authorities in favour of the contrary opinion, that some further consideration of the subject will not be misplaced.

The determining principle of value is the cost of procuring a similar service. When one man has a nugget of gold, or a bird of paradise, or any other precious object, if another desire to obtain from him this treasure, the owner will obviously reply to the objections of the intending purchaser that, if he dislike the offered terms, he may get another such treasure for himself. The purchaser will feel the force of this answer; and will conclude or abandon the bargain according to the intensity of his desire for the object, and his prospects of obtaining it on better terms elsewhere. In each of these cases, as in the case of the water which M. Bastiat illustrates, the subject of payment is not the natural agent, but the labour saved in its attainment. If there be no labour, there will be no price. If the labour be great or small, the price will vary accordingly. But whether the labour be great or small, whether the natural agents be abundant or scarce, the properties of these agents are unchanged; and the price is paid not for them but for the service rendered in procuring them. The subject matter of the exchange does not of course affect the nature of the transaction. The principle of price remains unaltered, although that subject matter may be not something above, or upon, or beneath, the earth, but a portion of that earth itself. When one man has a piece of land unusually fertile or well situated or otherwise suitable for human purposes, if another wish to obtain possession of that land, the reply of the landowner will still be the same as that of the owner of the chattel in similar circumstances, "Either accept my terms, or get another for yourself." If the purchaser think that he can upon the whole obtain at a less cost another piece of land equally convenient, he will take the latter part of the alternative. If on the contrary he think that he cannot suit himself equally well elsewhere, or that if he can so suit himself the cost will on the whole be greater than the price demanded, he will pay that price. In this case, as in the case of the gold or the bird or the water, the purchaser pays for the service rendered to him; for the trouble he is saved, and for the convenience of having then and there the object that he wants.

§ 6. Land therefore, like other natural agents, commands a price then, and only then, when it has been in some manner affected by human labour. In most cases the presence of this additional element is sufficiently evident. Saleable land has generally been improved; or is in the vicinity of markets, or of roads, or of other objects on which capital has been expended. Men prefer to pay for land so circumstanced than to submit to the delay and trouble of bringing the wilderness into a similar condition. Sometimes when suitable land is not easily discoverable, they prefer to pay for its discovery than to undergo the risk and fatigue of a search. But there are cases in which mere waste land, far from any improvement, will, even from the first occupier, command a price. In America and Australia, no land is now alienated to private persons except by auction or for a certain minimum price fixed by law. This sum is not regarded as a mere class tax, but is cheerfully paid by purchasers as the fair equivalent of a benefit received. If this price be not paid for the inherent powers of the natural agent, for what is it paid?

In reply to this question it may be first observed that in these countries tens of thousands of acres, and these not sterile but of average and often more than average fertility, though duly offered for sale, frequently remain for years without a purchaser. The lands that are taken up are those which, if not improved, are at least within the limits, so to speak, of improvements; which are near other settled lands, and are likely to participate in the benefits of such settlement. But even in the case of the first purchaser in a new district, the purchase is not without consideration. The state guarantees to him his heirs and assigns for ever the peaceful and undisturbed possession of the land he has chosen. If he wanted cheap land, he might have found abundance in Central Africa; most persons however prefer to pay one pound an acre for poor land in Victoria, than to take up at their will an unlimited extent of fertile country on the banks of Lake Chad. But even without leaving his own country, a new settler might in most cases take unauthorized possession of the land that we assume him to purchase. If he do so, he is of course free from any charge;

but he acts at his peril. The natural agents in any country are of common right. Every inhabitant has an equal title to their use. The property in them, if I may so speak, is undivided. To avoid confusion, therefore, every community claims the right of determining the conditions on which the appropriation of at least the most important of these agents, or of those of which the use is continuous, may take place. In the case of land some grant from the sovereign, or other competent authority representing the public, has always been required to confer upon the occupant an exclusive right; and to bar the claims of all other persons who by virtue of their nationality might claim an equal interest. If such a grant be made the subject of pecuniary consideration, it is because the grantee prefers to pay the price demanded than to run the risk of dispossession. Nor is this apparent sale and purchase of a natural agent confined to land. The state also issues for value licences to cut its woods, to remove its gold, to consume its grass. The property of all these objects is in the state or collective community; the service which the state renders, and for which payment to it is made, is the extinguishment of all other rights, and the security given to the purchaser. Men therefore pay not for the actual land, but for the title to that land. Accordingly we see that when they can, free of charge, occupy land without a title, they are yet willing to pay for a grant in fee of that very land, in the expectation that this grant will permanently secure their undisturbed possession.

§ 7. I have already said that the ultimate partners in any production may be divided into two classes, capitalists and labourers. Since every product implies labour exerted upon natural agents, and aided by capital, invention, co-operation, and exchange, or by some at least of these agencies, and since the co-operation of nature, whether its character be apparently passive or energetic and active, is gratuitous, and since exchange is only indirectly connected with production, the only remaining parties are the labourer, the capitalist, and the co-operator. Of these the first and the third are of the same kind ; and we may therefore confine our attention to the capitalist and the labourer, whether the latter be the originator of the work, or be merely a co-operator in its performance. Our present inquiry relates to the circumstances which determine the respective shares of these partners in the result of their combined industry. On no economic question has more been written than on the relations between these two classes ; and on no such question have the discussions been less satisfactory. It will be necessary therefore carefully to distinguish between the different subjects involved in these discussions, and to ascertain precisely the meanings of the terms of most frequent use in them.

Where the number of partners is not inconveniently great, and where the practice of express co-operation is familiar, the proceeds are divided between all the partners according to the terms of their agreement. All share alike in the profits, and no distinction is made between the different classes. In other circumstances the property in the product, although it is divisible between the capitalist and the labourers, is usually vested in one party only who distributes their share to the others. The residue in the hands of the owner, after paying to all his partners their several shares, is called profit. This term is usually applied by economists to the remuneration of capital; but since when used in this sense it includes the wages of superintendence and the insurance against risk, it may without impropriety be used as equivalent to *gain*. If the distributor be the capitalist, the share of the labourer is called wages. If the distributor be the labourer, the share of the capitalist is called either interest or rent. It is called interest when the capital has been advanced in the form of money. It is called rent when the capital has been advanced in the form of some specific commodity. Thus rent is paid upon a house or a machine ; interest is paid upon a mortgage or other loan. Whether the distributor be the capitalist or the labourer, the shares of the co-operators will be wages.

We have seen that capital is a condition precedent to the use of invention. The forces of nature, although they require no reward for their work, yet claim before that work is performed the fulfilment of certain costly conditions. But when this aid has been obtained, the efforts of the labourer are rendered far more efficient than when they were unaided. In the early days of a worker or of a nation of workers, capital is scarce, and obtained only with difficulty In such circumstances its assistance is a service of the utmost importance ; and the labourer is willing to concede to its proprietor a large share of the product. Yet although its share is thus great, the aid which capital in such a case renders is comparatively small ; and the natural forces which yield to its puny efforts are the least potent of their kind. The product is consequently small ; and the receipts both of the capitalist with his large share, and of the labourer with his small share, are alike insignificant. Still the product, humble though it be, is much larger than it would have been without capital and invention ; and thus an additional accumulation from the increased production becomes possible. Assuming this accumulation to take place and the same process to be repeated, capital will gradually become more plentiful, and so more attainable, and so less dear. The service now rendered in supplying it is less than that rendered when its attainment was difficult. Capital therefore commands, as it increases, a smaller share of the product than in the times of scanty accumulations it could obtain.

This tendency however towards a continued diminution of its share is speedily checked. The increase of capital implies gain. If there were no gain, there would be no motive for investment. Consequently the two conditions of an increased capital, the reduction of interest and the existence of gain, must be consistent. But although the capitalist's share of the product is thus diminished, yet as the product itself continues to increase, the smaller share of the larger product is more than equivalent to the larger share of the smaller product. Five per cent. on three hundred pounds gives a larger return than ten per cent. on one hundred. As the share of the capitalist decreases, that of the labourer necessarily increases; so that ultimately the latter has both a large share of the product, and receives a large sum as the proceeds of this share. Nor has the capitalist any reason to be dissatisfied with this result; for he in like manner, although his proportion of the gross product is small, receives a larger sum than he formerly did. Abundance therefore supplies ever increasing products; and large products yield to all parties interested in them large receipts. Scarcity on the contrary gives small products, and consequently small receipts. Thus it is that wealth begets wealth, while the struggle to emerge from poverty is protracted and hard. In the former case, the wealthy capitalist is little troubled by the smallness of his share: in the latter case, the greatness of its proportion is but a scanty consolation to poverty. It is manifestly the amount of the receipts, and not their relation to the product, that is the real object of interest to all the partners.

§ 8. In that tacit partnership however which society forms, a new arrangement practically supersedes this division of shares. The owner, or, as we might call him, the managing partner in this virtual firm, purchases, as we have seen, the yet unrealized interests of his partners. To effect this object capital is of course required; but the actual owner of that capital is not necessarily the person who advances it. A man may borrow money at interest for some undertaking, and may pay labourers to help him. In such circumstances he discounts the shares of his partner the capitalist, and of his partners the labourers. He takes on himself the whole responsibility of the undertaking. If he gain, he enjoys the whole benefit. If he lose, he bears the whole loss. His partners have preferred a certainty to a contingency, an immediate to a distant return, a sure and moderate gain to the chance of a great prize or of an entire loss. Thus in every such case there are two distinct transactions: there is a partnership, and there is a mode of dealing with the shares of the partnership. In the latter point of view all questions as to the distribution of shares become practically unimportant. The parties look each to the amount that he actually receives, and not to his unascertained share of a contingent product. For their immediate purposes the question as to the terms of co-operation merges into an ordinary case of exchange. The relation between the parties is no longer that of partners, but that of vendor and purchaser. In place of having a share in the undertaking, the co-operator sells for a stipulated price his labour or the use of his capital. The case therefore comes within the ordinary conditions of exchange; and the price of labour and the price of capital are determined in the same manner as all other questions of price are determined. Yet the general character of the partnership is not destroyed. Although each particular transaction amounts to a sale, yet for the continuance of the business a nearer connection arises. Although the whole loss of the undertaking, if the undertaking be unfortunate, falls upon the last proprietor, and the interests of the other parties have been previously secured, yet each such loss prevents a repetition of the transaction from which it arose. The capital which ought to have been replaced, and which if replaced would have afforded the means of employing labour and of defraying the interest upon other capital, has disappeared; and thus the market for labour and for capital is by so much diminished. Both the labourer and the intermediate capitalist are therefore directly concerned in the success of every enterprise towards which they have contributed. If it be successful, they feel the advantage; if it be not successful, they feel in like manner the loss. But this community of interest is no longer direct, but is indirect merely; and it arises not from the gains or the losses of partners, but from the increased ability, or the diminished demands, of customers.

§ 9. There is a further question which in treating of wages economists have generally thought it necessary to discuss. It relates not to the labourer's share of the product, or to the amount of money which his labour realizes; but to his real wages as they are termed, or the commodities which his labour enables him to purchase. We thus regard the man not as the vendor of labour but as the purchaser of commodities. We follow, if I may use the expression, the purchase money of his labour into the labourer's hands, and make inquiries as to its application. But such inquiries, however interesting and important they may be, belong rather to the practical statesman than to the economist. There is no greater reason why for scientific purposes such an inquiry should be undertaken in the case of the labourer than in the case of the capitalist or of any vendor. Assuming that money is the standard of exchange, and that this standard is constant, we are concerned with the price, and the price only. The price of labour depends upon the laws of exchange as applied to labour. The price of the commodities which the labourer uses depends upon the laws of exchange as applied to each such commodity. The condition of the labourer depends upon the price of his labour, and upon the price of those objects, whether commodities or not, which he desires. In other words the labourer, like every other man, is " rich or poor according to the degree in which he can afford to enjoy (either by his own efforts or by the efforts of others) the necessaries, conveniences, and amusements of human life." The question whether in any particular community there be any disturbing circumstances that affect the purchasing power of any class in that community, or the objects upon which that power would if unimpeded be exercised, is one that depends upon special facts; and does not belong to the general theory of industry.

CHAPTER XIX.

OF COMPETITION.

§ 1. We have seen that in the social state exchange by substituting for an express agreement a reasonable and provident anticipation of men's mutual wants produces the effect of a tacit co-operation. As it is by a form of exchange that even in cases of express co-operation a distribution of the produce is made, it might be expected that in those forms of co-operation which chiefly depend upon exchange the distributive function of that agency would be prominent. Such is accordingly the case, although the social state introduces into the principle of exchange some remarkable modifications. In every exchange the ruling element of price is the difficulty of attainment, or in other words the cost of procuring a service similar to that which is the subject of the bargain. But as the rendering of a service implies gain to the party who renders it, if in any particular case the gain from any cause be unusually great or unusually small, if the service rendered receive an unusual remuneration, either in excess or in defect, the same motives and the same foresight which under the influence of exchange led to the growth and the permanence of separate industrial classes, will in a similar state of society induce several persons to undertake or to avoid, as the case may be, the performance of this service. The number of the persons thus induced to act or to forbear is determined by the amount of the remuneration; and this amount in turn is measured by the intensity of the want. Where there is the hope of gain, each competitor, whether the competition be among vendors or among purchasers, will be content with a smaller share of the surplus advantage than his predecessors or his neighbors. Where there is risk of loss, each person otherwise likely to perform the service will in like manner be reluctant to perform it. This process continues as long as the prospect of any unusual gain or unusual risk remains; until by degrees the whole advantage is exhausted, or the whole risk compensated, and the service is at length rendered at the usual rate.

When a system of free exchange is established, this competition is rather potential than actual, but is not on that account the less effective. The demands of each party are kept in check by the knowledge that there are others who either are or soon will be prepared to render on at least equally favourable terms the required service. Thus the purchaser is the gainer by the difference between the extreme amount which his desire for the service would induce him to give, and the amount actually paid. But as exchange implies gain not to one of the parties only but to both, the limit to this reduction of price in favour of the purchaser must be the point at which the performance of the service remunerates the vendor. That remuneration manifestly can never permanently fall below the actual burden to him of rendering the service, and under competition exceeds the amount of that burden only within fixed limits. This central point to which all the oscillations of price continually return is termed the cost of production, a phrase which includes not only the actual outlay of the vendor, but his own remuneration measured by the average reward then and there obtainable by similar kinds of labour. Thus from the conflict of various interests, there is evolved a certain price which at that time and that place prevails, and which represents the difficulty of then and there attaining the service in question. In the social state therefore price tends towards the cost of production.

§ 2. This readiness on the part of purchasers to outbid, on the part of sellers to undersell one another, is called competition. But although it thus bears a positive name, it must not be regarded as some separate entity. Competition merely implies the absence of restraint. It expresses the common right of every man to exchange his property or his labour on the terms which he thinks expedient. It expresses the influence exercised by the self-regard of one man upon the self-regard of another. It arises from the right to pursue a man's own interest, subject only to the limitation of not obstructing his neighbor's corresponding right. Thus every interference with competition is an interference with our natural freedom of action; and every restriction on competition must be justified in the same manner as any other restriction upon our liberty. As liberty is evolved out of the limitation of individual rights that the rights of other persons produce; so competition, which is merely a particular case of liberty, is evolved from that modification of self-interest to which the conflicting self-interests of others give rise. But this modification is attended with important and unexpected results. Under the influence of competition we see self-regard always realizing what it always is striving to avoid, and always failing in that which it is most anxious to attain. It constantly seeks its own gratification ; and it constantly extends the gratification of others. It constantly strives to appropriate the bounty of nature ; and it is constantly compelled to bring its acquisitions into the stock of common enjoyments. Men's interest, it has been said, acts as a spur to themselves, and as a bit to their neighbours; and the whole result incessantly tends towards the general advantage and elevation of our kind.

§ 3. In the absence then of any disturbing circumstances, and assuming that the principle of self-regard is in due activity, the same kind and amount of service, including in that term both commodities and services usually so called, cannot in the same market and at the same time command two prices. If one man ask too much, another will be more moderate. If one man offer too little, another will be a better customer. Thus the amount of effort required for the performance of any service is ascertained; and for that amount only payment is made. In the case of the same class of services, this principle is sufficiently plain. But though it may not be equally apparent, the principle is not less true as between different classes. In any community where competition prevails, the remuneration of all classes of services tends towards an equality. The number of those who are prepared to render the several services will adjust themselves so that the balance may be corrected. More or fewer persons, as the case may be, will engage in the business. More or less capital will be employed in it. "The numerous and multifarious channels of credit, through which in commercial nations unemployed capital diffuses itself over the field of employment, flowing over in greater abundance to the lower levels, are the means by which the equalization is accomplished. The process consists in a limitation by one class of dealers or producers, and an extension by the other of that portion of their business which is carried on with borrowed capital."* If the business be thriving, those who are engaged in it not only invest in it

---

* Mill's *Political Economy*, vol. i. p. 493.

all their own disposable funds; but in consequence of its prosperity readily obtain, and employ in its extension, advances from other persons. If it become unprofitable, the capital employed in it will not be replaced; borrowing for its use will cease, and the operations of the business will be restricted. It is true that great and obvious differences between the actual remuneration of various employments actually exist. Knowledge and practical skill can be acquired only by a great cost of time and trouble, and are consequently possessed but by few. Moral qualities, and the habits and sentiments which careful training give, are also rare. The services in which these qualifications are required are therefore greater, and consequently more highly remunerated than other services. There are also causes which in some cases reduce the amount of remuneration. Every man, as Adam Smith has remarked, has an undue confidence, if not in his own abilities, at least in his own good fortune, and over-estimates his chances of success. In early life especially, hope prevails over reason. Accordingly, where any employment contains a few splendid prizes, even though its average rewards be very low, it will attract many candidates. Men will often commute for a certainty their chances of a large but precarious income. Money, too, is not the only object that men seek: and consequently is not the only form that remuneration may take. A position of but moderate emolument may bring with it reputation, or power, or the opportunity of pursuing some favourite object. Such places are often sought with extraordinary eagerness. The remuneration is really high, if it be estimated as composed partly of money and partly of other considerations. But these cases do not affect the principle that equal services tend to a uniformity of price. They merely show the circumstances which must be taken into account in determining what services are equal.

§ 4. Thus equal efforts fairly interchange. The man whose want is supplied by another man returns to that other, as in justice he ought to return, an effort corresponding to that which was made in his behalf. But this equality is soon disturbed. Though there cannot be two prices in the same market for the same service, there may be two costs of production. Different persons may from a variety of causes be able to render the same service at very different degrees of trouble to themselves. In such circumstances, as there is but one price, that price will from the competition of the buyers be determined by the cost of production of the dearest part. The other producers will therefore gain, in addition to their ordinary remuneration, the difference between the market price and their respective costs of production. This extraordinary gain at once leads to and is corrected by competition. Whatever may have been the conduct which led to the reduced cost of production, other men quickly discover and imitate it; and the price from the competition of the new comers gradually falls, until the market price becomes that of the lower, not of the higher, cost of production. Thus competition secures both steadiness of price and uniformity of profit. The accidental advantages last so long only as their presence is necessary to excite imitation. When this object has been fully accomplished, equal efforts once more interchange.

Since the price rises or falls according to the intensity of the want, and since the number of competitors increases or decreases according to the rise in the price, each man is paid according to the merit of the service that he performs. The test of the importance of any service is the price which the recipient is willing to pay for it. Its importance is always relative to his desires. A song is a greater service to a wealthy man than a loaf of bread. The bread is indeed essential to his existence, while the music is a mere luxury;

but he can easily obtain the former, while the latter, at least when it is of a certain quality, is very rare. Accordingly the service of the vocalist is in the circumstances greater, and is therefore more largely remunerated, than that of the baker. It is not inconsistent with natural justice that the person who has done much should receive much ; and that the person who has done little should receive little. What is much and what is little, every man must judge for himself. A man's estimate may be and too often is morally wrong ; but his inability to appreciate things at their true worth does not affect his sense of the service done to him, or in other words of the assistance he has received in gratifying his desire, whatever that may be. But in this tendency to a fall in price, even the favoured producer, whose advantages gradually melt away, sustains no loss. He is remitted to the original conditions of production. He no longer makes any profit beyond his neighbors ; but he makes the same profit that they make. His increased reward was temporary, because his increased service was temporary. The service that he renders costs him a certain effort ; and a corresponding effort and no more is now returned to him.

While the producer has thus no reasonable ground of complaint, the benefit to the consumer is very great. A connection is now established between the service and its price. A check is thereby imposed upon the greed of the vendor. His power of running up the price to the extreme limit of the purchaser's desires irrespective of the cost to himself is defeated. The difficulty of attaining the object, not its desirability, becomes the immediate regulator of the price. Thus while the producer is remunerated according to the full measure of his deserts, the consumer gains the whole difference between the price that he actually pays and the price that in extremity he would be prepared to pay. The principle of competition then may fairly claim to be described as beneficent, just, and equalizing. It is beneficent, because it enables each man to supply his wants at a price determined not by the amount of the convenience which he receives, but by the amount of trouble which the performance of the service necessarily involves. It is just, for it rewards each man for and in proportion to his actual merit. It is equalizing, for it does not permit any man permanently to retain any advantage above other men, which he may chance to have acquired.

§ 5. This equalizing influence of competition is widely different from that artificial levelling which some thinkers have sought to attain. Unlike their systems, this tendency gives free scope to the action of self-interest. It does not destroy, it regulates, that great source of social activity. Its regulation is such that while self-interest is incessantly occupied in the pursuit of gain, every gain that it acquires quickly passes from individual appropriation into the domain of common right. Self-interest discovers and renders available for human purposes the gifts of nature. Competition rescues the gifts thus discovered from individual greed, and restores them as the free bounty of Heaven to all the children of men. Natural agents of various kinds and in various quantities are scattered over the whole earth. Yet no nation monopolizes its own physical resources. If one country have the exclusive possession of an article in general demand, wages and profits in that country will rise; the increased remuneration thus given to labour and to capital will attract from other countries labourers and capitalists; or at least the conditions will be favourable to the natural increase of labour and of capital. Wages and profits will consequently fall, until at length they are on the same scale as in other countries. The price of the monopolized commodity will then remain at the permanent and decreased cost of production. In other words the normal state of production will have been established. Consumers will pay for human agency and for that only. The peculiar bounty of nature will be as free to them, that is to all the rest of the world, as it is to the inhabitants of the favoured country itself. In the case of discoveries and inventions, the operation of the same influence is so clearly seen that even at the risk of repetition they deserve a special notice.

Natural forces may often be so directed as to diminish the labour of man. We have seen that price constantly gravitates towards the quantity of labour required for the production. If one man then can reduce that quantity of labour, while with other men the amount of labour remains unchanged, he will gain the entire difference between his reduced cost of production and the ordinary cost. A strong motive for invention is thus supplied, and men are induced anxiously to search for the means by which the powers of nature may be forced to do the work previously performed by human muscles. When such means are found, the inventor has a peculiar advantage which he can either use himself, or for the communication of which he can obtain from others a consideration. But such knowledge never remains long secret. The nature of the process becomes gradually known, and other persons begin to adopt it. As the number of these imitators increases, the price falls. At length the knowledge of the invention is generally diffused; competition reduces wages and profits to the ordinary level in other occupations, and the price returns to the cost, but now the diminished cost, of production. Through these three stages, invention, imitation, diffusion, to use the language of M. Bastiat, every process which substitutes inanimate forces for muscular action passes. A premium, self-adjusted according to the merit, is given to the successful application of a knowledge of nature to the purposes of life.

Smaller but still ample rewards are given to those who are the first to bring into general use these inventions. These rewards gradually diminish as the service becomes less meritorious; but still continue to some extent until the final result of universal diffusion is secured; and mankind obtain in the diminished sacrifices which their enjoyments demand the full and entire benefit. "I confess," says M. Bastiat,* "that the wisdom and the beauty of these laws strike me with admiration and respect. I see in them St. Simonism, *to each according to his capacity, to each capacity according to its works.* I see in them communism, that is to say the tendency of natural advantages to become the common heritage of man; but a St. Simonism, a communism, regulated by infinite foresight, and not abandoned to the weakness to the passions and to the arbitrary will of men."

Thus then under the combined action of self-interest and of competition, that is from the pursuit by many persons of their respective interests, there arises a constant tendency to diminish the sacrifices required for enjoyment, and that not for one but for all. There is a constant tendency to set the best intellects of the world to work primarily indeed for their own, but ultimately for the general, benefit. There is also a constant tendency to release men from merely mechanical work, and to devolve upon them labour requiring intelligence and skill. Such at least is the natural order which Providence has established, and which, however impeded by human folly, still upon the whole prevails. Human labour always requires an equivalent; but the aid which nature gives is free and priceless, continually relieving man from his severest toils, continually cheapening and increasing his enjoyments, and continually tending to equalize labour as far as the differences of intelligence and of moral qualities will admit.

---

* *Popular Fallacies,* p. 67.

§ 6. It is not our gains only that competition distributes: it distributes also our losses. As it effects the former result by lowering price, so it effects the latter by raising it. If from any cause the quantity of any article of consumption be diminished, the difficulty of procuring that article is thereby increased. But as its desirability is unchanged, the competition of the purchasers will continue; and the price of the remaining portion of the article will consequently rise. The effect of this rise in price is to diminish the consumption of the commodity. It places outside the conditions of exchange a number of persons who were previously within these conditions. Fewer persons use the commodity than before its scarcity; and the quantity of it which even those who continue to use it consume, is in many instances considerably reduced. Thus partly by the diminished number of consumers and partly by the diminished quantity that many at least of them use, the consumption is adjusted to the existing quantity. In cases therefore of diminished production, competition through its action upon prices operates as an insurance. It spreads over a large surface with a comparatively slight inconvenience a loss that if concentrated upon a particular class would be ruinous. If a portion of the crop be lost, the producers are to some extent compensated by the high price of the remainder. A part, indeed, of the loss falls upon them; but its principal portion is borne by the consumers, that is by the general public, in the form of higher prices. Thus no damage can be sustained by, and no wrong inflicted upon, one part of the community exclusively. If no man can secure a strict monopoly of the good that he possesses, so no man can ensure to himself complete exemption from the harm that befalls others. Nor is the reason of this extended liability obscure. Not only does society thus become a kind of great mutual insurance company; but it gives as it were security for its good behaviour towards each of its members, and in turn receives a similar security from them. The principle operates both to secure from total loss the victim of accidental calamity, and to guard individuals against the tyranny of a majority, and at the same time to bring public opinion, where positive law is defective or inappropriate, to bear upon the conduct of each member of the community. If the majority enact an unjust and oppressive law, they soon find that the injury it occasions extends to themselves. If a citizen waste his substance in riotous living, he is regarded by prudent men not merely as unwise but as a public nuisance. " The inevitable sanction of an exact distribution of justice addresses itself to men's interests, enlightens opinion, proclaims and establishes among men those maxims of eternal truth, that the useful is one of the aspects of the just, that liberty is the fairest of social harmonies, and that honesty is the best policy."*

§ 7. Another result of competition is that it secures for every undertaking the most efficient administration. It is the activity of interest quickened by emulation that gives its surpassing energy to private and unassisted enterprise. The real and effectual discipline which is exercised over a workman is, as Adam Smith remarks, not that of his corporation but of his customers. It is the fear of losing his employment that, even when conscience is silent and law asleep, restrains his frauds and corrects his negligence. The antiseptic influence of free competition is too well known to need any copious illustration. Few persons at the present day will, at least as regards domestic industry, dissent from Gibbon's observation that " the spirit of monopolists is barren, lazy, and oppressive : their work is more costly and less productive than that of independent artists and the new

---

* Bastiat's *Harmonies*, p. 287.

improvements so eagerly grasped by the competition of freedom are admitted with slow and sullen reluctance in those proud corporations above the fear of a rival and below the confession of an error." But the unfailing Nemesis of outraged nature is sure to overtake all such breaches of her laws. Special privileges, whether they be industrial or political, are even more injurious to those whom they are meant to benefit than those whom they willfully injure. Under their influence the natural discipline of the workman is weakened; and consequently his labour becomes less productive, and his character is deteriorated. He becomes not only indolent and apathetic, but violent and jealous. He relies not upon himself but upon his monopoly, and fiercely resents even the slightest infringement upon its extent. Yet it is often found that when he is forced to part with his exclusive privilege, new life is quickly infused into his business. His services are more widely demanded than before; his production more abundant; his receipts more ample.

Even in international relations, monopoly so far from benefiting actually injures the favoured country. Of this truth the experience of the present reign furnishes many remarkable examples. For the space of nearly twenty years prior to the introduction of free trade, British exports remained stationary; but upon the change of policy they began to increase, and advanced with each extension of free trade principles.* The rice of Bengal, to give a more specific instance, was at one time considered so inferior to that of America that even with the aid of a protecting duty of three halfpence in the pound, it could hardly compete with the latter in the English markets. When this duty was first reduced in 1842, the complete and immediate annihilation of the rice

---

* *Journal of Statistical Society*, vol. xviii. p. 173.

trade was predicted. To avert this evil an intelligent and enterprising merchant forthwith took measures for improving the quality and appearance of the shipments from Bengal.* The result of the exertions thus made has been that so far from the India rice trade being destroyed it has increased 635 per cent. upon its extent in the good old days, thirty years gone by, of protective duties. The Indian imports are much more than sufficient to meet the demands of the English home markets, and the continental markets receive a large portion of their rice from England.† That nervous apprehension and restless jealousy characteristic of a system that depends or is supposed to depend upon external support is well illustrated by a curious incident in our commercial history. In 1778 among other proposed relaxations of the commercial laws in favor of Ireland, a bill was introduced into the English Parliament to permit the importation into England of Irish sail-cloth. This measure was violently opposed, and petitions were presented from all parts of the kingdom setting forth in strong terms the ruinous consequences that must follow the removal of its natural protection from this important branch of British industry. It was said that in Ireland taxes were low and labour cheap, and that consequently the Irish would be able to undersell the English manufacturers to such a degree that several of the principle seats of trade would inevitably be ruined. These prophetic warnings would certainly have been fulfilled, had it not been unfortunately discovered a short time afterwards that the liberty of importing Irish sail-cloth had been already established by a positive law of long standing.‡

§ 8. But competition has another function. By its

---

* Porter's *Progress of the Nation*, p. 744.
† *Journal of Statistical Society*, vol. xxiii. p. 69.
‡ *Annual Register*, 1778, p. 177.

means the principle of natural selection is applied to industry. As in nature various existences struggle together and as it were compete either with each other or with the circumstances of their position, until none but the strongest and healthiest of them survive and continue their race; so in society a similar process is in constant operation. If the weaker grasses be killed by the stronger ones, so the feeble or unskilful tradesman falls before his superior competitor. It is not more certain that one species of rats or of cockroaches has frequently expelled another such species, or that in Yorkshire the long-horns displaced the old black cattle, and were themselves exterminated by the short-horns,* than it is that the hand-loom weavers are disappearing before the power-looms, or that the old stage coachman has given way to the stoker. Competition in like manner prevents the rise or rather the continued existence of unsuitable projects. It nips in the bud the thousand schemes which, though not absolutely impracticable, are unprofitable or premature. It gauges the intensity of the social wants, and provides for the proper order of their satisfaction. In this process some loss is inevitable. Until hard experience has taught them, men will miscalculate both the desires that they ought to supply, and the extent of the supply that is required. It is the function of competition to ascertain these facts, and to enforce attention to them. It may be that in many cases ten capitals are used where nine would be sufficient. But there is no other means save competition of determining that nine are enough and not more than enough; that eight are too few, and ten too many. Valuation, whatever form it may assume can never supersede competition, because it is the competition itself which establishes the standard by which the valuators must proceed, and so renders the valua-

---

* Darwin's *Origin of Species*, pp. 76. 111.

tion possible. But such waste, if waste it be, is equally observable in nature. Millions of seeds fail to germinate, tens of millions of seedlings perish, while comparatively few attain perfection. What death does in nature, insolvency effects in society. Unsuitable undertakings, or unsuitable persons for the conduct of promising undertakings, are infallibly detected, and relentlessly removed. By this stern yet salutary law the ultimate benefit of the species whether physical or social is secured. Of society, as well as of organic nature, " it may metaphorically be said that natural selection is daily and hourly scrutinizing, throughout the world, every variation even the slightest; rejecting that which is bad, preserving and adding up all that is good, silently and insensibly working whenever and wherever opportunity offers at the improvement of each organic being in relation to its organic and inorganic conditions of life."* There is however one conspicuous difference between the two cases. With natural organisms such repression is always necessary. But in society its purpose is but temporary. It serves to extirpate ignorance and mistakes and blunders. But these shortcomings are not necessary to man. That stern rugged nurse's rigid lore is designed to teach men to avoid these errors and to train them to a better way. The penalty is sharp; its exaction is unfailing; yet hitherto the pupil's progress has been but slow.

---

* Darwin's *Origin of Species*, p. 84.

CHAPTER XX.

OF THE SOCIAL CONTRIVANCES TO PROMOTE ORGANIZATION.

§ 1. Associated industry with all its advantages has its own cost. Its great advantage springs from its division of employments. But this division depends upon the extent and the state of the market, that is, upon the power of easily and quickly effecting a sale. The completion of a sale however is often a slow and difficult process. Purchasers must be found; terms must be settled; and goods must be delivered. All these proceedings involve in a greater or less degree the expenditure of time and of trouble. This expenditure is in fact the price we pay for the advantages of the association. The advantages may more than outweigh the price; yet still its pressure is severely felt. Various expedients have been therefore contrived to facilitate exchange, and to reduce the influence of its characteristic imperfection. Since these expedients relate to social phenomena, they are found only in a state of society. They spring from an application of the associative principle for the purpose of its own subsequent extension. They are in effect secondary industrial aids. They do not directly aid industry; but they are contrivances to promote the action of two primary auxiliaries, co-operation and exchange. Since the immediate difficulty arises out of the process of exchange, these arrangements are directly designed to facilitate that operation. They therefore relate either to the means of settling the terms of the exchange, or to the means of discovering new purchasers, or to the means of placing the purchaser when found in possession of the subject of the transaction. In other words they relate, if we use the terms in their widest sense, partly to sale and partly to delivery.

§ 2. One of the earliest and most pressing difficulties in cases of barter must have been the want of a common language of quantity. The numerals are among the very oldest of words; and the nations of Modern Europe and their descendents still retain under whatever disguises the same terms of numeration by the aid of which their Arian ancestors reckoned. But terms were wanting not for numbers only, but to describe the various quantities of capacity, of length, and of weight. The varying natural objects, such as a grain of corn, the human foot, or the compass of the human arms, which men seem to have first used as such measures, were soon felt to be too vague for even the earliest stage of commercial purposes. "Uniformity of weights and measures," says Bentham,* "under the same government, and under a people who in other respects have the same language, is a point upon which it would seem that there is no need of much reasoning to show its utility. A measure of which an individual does not know the contents is useless. If the measures of two towns are not the same either in name or quantity, the trade between the individuals cannot but be exposed to great mistakes or great difficulties. These two places in this respect are strangers one to another. If the nominal price of the goods measured be the same, and the measures are different, the real price is different; con-

---

* *Works*, vol. i. p. 555.

tinual attention is requisite, and distrust mingles with the course of affairs; errors glide into honest transactions, and fraud hides itself under deceptive denominations."

Some attempt has been made by almost every nation, and at an early period of its history, to obtain an exact definition of this important class of terms. But it is remarkable how inoperative in such cases is positive law. It is indeed, as Bentham observes, "a rare and noble thing to see a government labouring upon one of the essential bases of union among mankind." So far from these labours having established throughout the world a uniform standard, experience has shown that it is impossible, except by very gradual changes and very slow degrees, to secure even in a single and highly civilized nation that uniformity the advantages of which all men acknowledge. In the laws of the Anglo-Saxon kings, a century before the Conquest, there are provisions to secure uniformity of weights and measures; and Magna Charta contains a similar enactment. Various Acts of Parliament with the same object were subsequently passed, but the desired uniformity could never be accomplished. "So forcible," as Lord Coke complains, "is custom with the multitude." It was so recently as 1835 that after many fruitless attempts and by much careful preparation of the public mind it became possible to attain our present system. Even still beer is sold in Kent by the bushel, and flour by the gallon; and in other places butter and milk are sold by the yard, and other measures strange to the out-dweller are kept up in various places in England. In America notwithstanding that the decimal division of money into dollars and cents has long been established, the people are said habitually to buy and sell by the shilling and the sixpence.[*] In France at the very beginning of the great revolution,

---

[*] See *Parliamentary Remembrancer*, vol. ii. p. 80.

when it was a matter of high political importance to awaken a keener sense of national unity, and when the sweeping social and political changes that attended the downfall of the old régime seemed to give the fairest opportunity for such an alteration, the National Convention passed a decree for the uniformity of weights and measures. This decree was regarded as one of the greatest benefits to the country that the revolution could bestow ; and heavy penalties were attached to disobedience. Yet all the influence of every successive government, and all the intrinsic merits and simplicity of the system have not, even up to the present time, enforced throughout the provinces the use of the legal standards.*

§ 3. But it is not a standard of quantity merely that men require for conducting their bargains. They need also a standard of value. "As it is much easier," says Mr. Mill,† "to compare different lengths by expressing them in a common language of feet and inches, so it is much easier to compare value by means of a common language of pounds, shillings, and pence. In no other way can values be arranged one above another in a scale ; in no other way can a person conveniently calculate the sum of his possessions ; and it is easier to ascertain and remember the relations of many things to one thing than the innumerable cross relations with one another."

The same machinery which is thus useful in settling the terms of an exchange is no less useful in bringing together purchasers. The standard of value must always be a commodity, the qualities of which and its relations to most other known objects have been tolerably well ascertained. The transition therefore is not violent when the standard is

---

* Porter's *Progress of the Nation*, p. 347.
† *Political Economy*, vol. ii. p. 4.

also made to serve as a medium of exchange. Men receive it not because they immediately require that specific commodity, but because they are satisfied that they will have no difficulty in selling it again. The facility in finding a purchaser arises from the same reason which has led to the selection of the article as the standard of value. This reason is because the article is the commodity in most general demand in that society. Various commodities have been used for this purpose in various communities: cattle among pastoral nations; fish among those who live by fishing; tobacco among Virginian planters; nuggets among gold-diggers; pieces of cloth among the negroes; salt among the Abyssinians; bags of cacao among the Aztecs; but all these commodities agree amid all their variety in the common quality of being objects of general demand. Experience however soon showed that there was one class of commodities that combined in an eminent degree all the qualities desirable in the machinery of exchange. The precious metals have at all times and in all nations been objects of universal desire. Their beauty no age can wither; and their variety no custom can stale. They are durable beyond most materials. They can be divided into the minutest portions. Their quality is unchanged in whatever quarter of the world they are found. Few other commodities of equal value are of so small a bulk; few others therefore can with such facility be transported, or, if occasion should require it, be concealed. To crown all these merits, experience has proved that these metals possess the further quality of being less subject to fluctuation in value than any other commodity. Accordingly in every nation that could afford to possess them, gold and silver have been used both as the standard by which the value of all things is estimated, and as the medium by which all other things are interchanged.

§ 4. This use of the precious metals arose neither from positive law, nor from general speculations on their nature. It was the spontaneous and gradual development of the attempt to satisfy an immediate want. In this as in every other case of the kind, the practise preceded the theory. Man felt the want of a standard of value, and found that gold and silver were very convenient for the purposes of such a standard. At a later period when sufficient experience had been accumulated, and men began to reflect upon their conduct and to generalize their motives, the standard thus adopted with no care beyond the supply of the immediate need was found to be the one which the most skilful foresight would have selected. In like manner, the extension of the standard of value to the purposes of a circulating medium proved equally successful. In its former capacity, the precious metal enabled the negotiating parties easily to settle the terms of their exchange. In its latter capacity, it removed the chief difficulties in finding purchasers. Without its intervention the problem is to find two persons each of whom desired not only to sell his own property, but to purchase that precise property in that precise quantity and of that precise quality which the other party was enabled to offer. With its aid, half the difficulty was at once removed. It was then needful simply to find a suitable vendor; and all trouble as to his requirements and to the application, so to speak, of the purchase consideration disappeared. Still however the transaction was not free from difficulty. In its origin the exchange of labour or of goods for gold or silver was as complete a barter as the exchange of the same goods for any other commodity. It was necessary to ascertain both the quantity and the quality of the precious metal. The gold or the silver must be divided, and must be weighed, and must be assayed. It was not until credit came into action that it was possible to change the commodity into a mere instrument, and to develope the simple barter into the more complex process of a sale and a subsequent purchase. If all men could be satisfied that a given piece of metal contained a certain quantity of gold or of silver, and of a certain fineness, the process of exchange would be greatly facilitated. To afford this facility, the practice of coinage arose.

§ 5. The prerogative of coinage has long been regarded as "the choicest flower of the royal diadem." Originally however and apart from its abuse, it hardly merits such an encomium. It is strictly analogous in its real nature to the prerogative of regulating weights and measures. It merely defines each coin as a certain quantity of a certain metal of a certain fineness. Such a barren privilege however fell far short of the general idea of what the prerogative of coinage implied. That prerogative in its legal sense further involves the monopoly of issuing all such pieces of metal guaranteed as described. The monopoly is secured partly by penalties and partly by rendering the coins so issued and none other legal tenders for all pecuniary obligations. It is needless to say how profitable this privilege was at a time when sovereign princes were not ashamed to commit acts for which a private citizen would justly now incur some of the severest penalties of the law. In France during the feudal period every lord had the right of coining money; and unsparing issues of adulterated coins proceeded from these seigneurial mints. In not a few cases a compromise was made, and the vassals paid to their lord a triennial rent-charge in consideration of his forbearing to debase his money. When the power of the crown became sufficiently established, the King of France claimed the right of debasement as the peculiar privilege of the crown, and one of its choicest prerogatives.* At the present time

---

* Hallam's *Middle Ages*, vol. i. p. 205.

in our country this monopoly is probably convenient, and certainly harmless. But the monopoly is not in any way essential to the institution for facilitating exchange. What is of importance is that the denomination of coins should be defined; that a pound should be declared to consist of a certain fraction' of an ounce of a metal consisting of well ascertained proportions of pure gold and alloy. It is not material however to this definition whether the purity of the coins so defined should be guaranteed by the faith of individuals or of the crown. At the time at which the public guarantee would have been really useful, it was not always deserving of much confidence. When it has become perfectly reliable, there is an abundance of persons whose security would, as the circulation of their notes proves, be not less trusted. It is fit that government should define a foot measure, or the capacity of a gallon; but there would be small advantage in giving to a public department a monopoly of the manufacture of quart bottles or of foot rules.

§ 6. But with all its advantages, a circulating medium composed of either of the precious metals is soon found to have two serious defects. Although gold and silver are relatively to their value very portable, still their actual transport involves no inconsiderable trouble, delay, and risk. In India where silver was the only legal tender and neither gold coin nor bank notes were in circulation, this inconvenience was severely felt. It was stated in evidence to a Parliamentary Committee, that just before the rebellion it was the practice to convey a lac of rupees under a guard of a hundred soldiers.* The pressure of this inconvenience increases with the increase of business. The machinery too

---

* Levi's *Annals of British Legislation*, vol. vii. p. 197.

which is thus felt to be imperfect, is also felt to be very costly. The precious metals, as their name denotes, require much time and trouble for their attainment. It is not every nation that can afford to use such expensive machinery. There is no community so opulent as not to derive a perceptible relief from the substitution of an equally efficient but less costly medium. Mr. M'Culloch * estimates that assuming the ordinary rate of profit to be six per cent., and allowing for loss by wear and by casualties, the annual cost of maintaining a metallic currency of eighty millions would exceed five and a half millions. He also shows that the French, with a much smaller amount of business than that transacted in England, pay for their currency, which is principally metallic, not less than six millions a year.

Accordingly as society advances, various expedients are adopted for economizing the currency. These expedients appear under the different forms of credit, bank notes, promissory notes, and bills of exchange. Whatever may be their variety of external form, all these obligations agree in the one leading characteristic, that they are promises to pay accepted in place of actual payment from a confidence that the promise will be fulfilled. By these means and by the mutual operations of the chief professional money dealers, the facility of exchanging is enormously increased; and a great saving of expense is at the same time effected. It has been estimated that in 1839 at the London Clearing House exchanges to the amount of three millions daily were effected with about £200,000 in notes and gold. There is every reason to believe that at the present day these figures are greatly below the truth. In New York the daily clearances are said to average upwards of four millions sterling, while the amount of coin actually used is quite inconsiderable.† Thus partly

---

* *Treatise on Money*, p. 426.
† Tooke's *History of Prices*, vol. vi. p. 752.

from the increased skill which the division of employments affords, and partly from the increasing trustworthiness of the community, not merely a less expensive but a more efficient machinery of exchange is introduced: and in proportion to the decrease in the obstacles to exchange, the actual power of exchanging increases. This function however of credit is distinct from its influence upon the employment of capital. Credit, as we have seen, is able to fructify capital, to collect it into convenient quantities, and to place it in appropriate hands. The banks are the great modern agents for this important work. But the function of credit with which we are now concerned is different. It affects not the power of utilizing capital, but the power of exchanging; and it increases the latter by substituting for the precious metals a less costly and a more transferable medium.

§ 7. But there are other and more primitive means than money for bringing together vendor and purchaser. When they desired to sell their labour or their goods, men have naturally resorted to some public place, the street or some thoroughfare, where many persons would be likely to pass, some of whom might be disposed to purchase. In such circumstances some particular place from some real or supposed convenience becomes gradually more frequented than other places; and the advantage once obtained is from its very nature continually increased. In the same manner some particular time of the day and some particular day become known as favourable for business. These days are generally holidays, days on which men abstain from their ordinary work, and assemble. some for amusement, some for the observances of their religion, some for the transaction of business, and most from all these motives combined. When, as in early Europe was formerly the case, some religious observance was the proximate cause of meeting, these assemblages were held at some peculiarly sacred place, and as a natural consequence at some regular intervals of time. Thus grew up what in the case of ordinary meetings were called markets, and what when the assemblage was unusually large and was held at greater intervals, were called fairs. As society advanced, the religious observances were separated from the commercial contrivance. The history of this change may clearly be traced in England. There as elsewhere the early fairs were held on Sundays and holidays, generally under the protection of some patron saint. The people were brought together to hear divine service ; and subsequently business was transacted. But in the year 1285 a statute of Edward the First prohibited the holding of fairs or markets in churchyards. A century and a-half afterwards * it was enacted that all showing of goods and merchandise except necessary victuals in fairs and markets was to cease on the great festivals of the church, and on all Sundays except the four Sundays in harvest. Ultimately, in the reign of Charles II., the holding of fairs and markets for any purpose on any Sunday was altogether forbidden.

This custom of frequent assemblies was not likely in the early days of modern Europe to escape the vigilant rapacity of the feudal lords. As in so many other cases, the common right became the subject of a seigneurial privilege. The lord or, when he could, the king claimed the privilege of determining when and how fairs should be held, and of imposing certain tolls upon those who frequented them. So in our law it is the prerogative of the crown to establish fairs and markets, subject to the condition of not thereby injuring any similar grants already made, or presumed to be made. In

---

* See 13 Edw. I. c. 5; 27 H. VI. c. 5; 29 C. II. c. 7.

most cases the regulation of the markets belongs to the town in which they are held. Partly with a view to the improvement of the property, for such these markets were, and partly no doubt from more liberal motives, various advantages were given to sales in markets over similar sales in other places. Thus it is held that the sale of a chattel in market overt, that is in open market during the ordinary hours of business is good even against the rightful owner of the thing sold.

It may perhaps be interesting to observe how fairs were regarded in Peru. Under the paternal despotism of the Inca, no person was permitted to be idle. The physical condition of the country gave rise to very marked differences in the industry of its several provinces. The use of money was unknown. Hence it was absolutely necessary to bring together buyers and sellers; and for this purpose fairs were proclaimed by the Government. These fairs, we are told, also answered the purpose of giving a holiday to the industrious labourer. Thus in early Europe where the gentle influence of the church was ever at work to soften the labourer's hard lot, the holiday brought with it the fair. In Peru, the fair was the main object; and the benefit to the labourer, if indeed the Incas thought it a benefit, was merely accidental.

§ 8. Although fairs and markets are useful and in some cases even necessary for exchange, they are not without their peculiar disadvantages. They are held at intervals more or less distant. If they have the merit of bringing many people together, they have also the defect of taking away many people from their ordinary occupations. This defect became more conspicuous when the number of church holidays was diminished, and their observance grew less strict. They are indeed the natural growth of an early state of society, when population is scanty and sparse. Their use marks a period of social growth that is analogous to an inferior stage of organic development. In those animals whose vascular system though distinct is imperfect, there is no regular circulation in definite courses, but the direction of their currents is periodically changed. The blood in each part of the body flows for a time in one direction; then stops; and then flows in an opposite direction. It is only where the organization is tolerably complete that there are constant currents in definite directions. Thus too when the social development is advanced, the local and variable currents of fairs and markets are merged in the rapid and constant streams which the well-marked channels of communication supply.* The old languid and sluggish movement is quickened into the rapid and regular pulse of civilized life.

Accordingly the importance of fairs is at the present day greater in Germany than in England; and in Russia than in Germany. In one Russian government we read of thirty-seven large fairs; in another of nearly three hundred.† In the more developed countries, they either disappear, or linger without any industrial utility as places of amusement; or when they survive they are frequented by a different class of persons from those that in earlier times resorted to them. As population increases, men find it more convenient to employ an agent than personally to attend the fair; and such agency when there is a sufficient constituency becomes a lucrative occupation. As the means of communication are improved, this tendency becomes more pronounced. Thus in every town and village of Great Britain shops are now found which supply almost all

---

* See *Westminster Review*, vol. xvii. p. 118.
† Haxthausen's *Russian Empire*, vol. i. pp. 171, 410.

the requirements of their districts, and by the aid of roads and railways and other modes of transit derive their own supplies from remote and opposite quarters. Fairs and markets in short give way before, or exist mainly for, a new class that is gradually developed. The great distributing class under its various forms of merchants, agents, factors, brokers, or other middlemen and retailers, furnish a separate series of occupations. The facilities for exchange which the growth of such a body offers are obvious. Those who wish to buy can get the article they want, and the quantity they want, at any moment and without leaving their own immediate locality. Those who wish to sell have no difficulty in finding either a purchaser or a person who can find a purchaser. Every shop and every house of business becomes in fact, what every shop in London is held for legal purposes to be, a market overt.

As society advances, these distributive agencies, like all other occupations, become more and more fully developed. But this progress is not unfrequently regarded with disfavour; and strange errors respecting it are even still prevalent in unexpected places. There is still a common belief that these middlemen, a term which is almost always dyslogistic, profit by other men's labour; that they are the drones of the industrial hive; that they produce nothing; but that they by their charges greatly add to the cost of production. Still more bitter is the feeling with which even yet corn merchants are regarded as persons who like the usurers derive their wealth from the misery and the necessities of other men. These views have indeed long been discarded by economists, and have even ceased to influence legislation; yet the importance of the function which the distributing class performs seems not so universally recognized as to render further illustration superfluous. Two opposite examples may be cited, the one exhibiting the efficiency of the natural agency; the other showing the difficulties to which the absence of that agency gives rise. I have already observed how, under the system of distribution that has spontaneously grown up, and with the assistance of an almost perfect system of communication, nearly three millions of people are fed in London in a manner which probably no equal number in the world enjoys, without any necessity for any extensive storage of provisions, and without the least apprehension of any failure in the supply, or even of any variation in the price from day to day in consequence of any shortcoming. It is instructive to contrast with this silent and unerring apparatus the pernicious process by which the population of ancient Rome was maintained. With unlimited wealth, with unlimited power, with the harvests of Sicily and of Africa at their command, with Egypt exclusively appropriated as a sort of granary for Rome, the masters of the Roman world were with difficulty able to provide for a quarter of a million of people their scanty dole of corn. I do not now speak of the ruin thus caused to Italian agriculture, or the demoralization of the urban populace, or the other political consequences of this interference of the state. It is enough to notice the difficulty and the danger with which this imperfect supply of a single article of food to a comparatively small population was attended. So terrible was the responsibility that it is said that the casual delay of some of his corn ships drove the cold and wary Augustus to the verge of suicide.*

Our own recent history presents a still more remarkable example. The experiment of an artificial system of distribution has been tried in our own times on the grandest scale, and its results are stated by unsuspected witnesses. In Ireland, during the great famine, in the month of March, and

---

* See Merivale's *History of the Romans*, vol. iii. p. 501.

again in July, 1847, according to official returns, the wonderful spectacle was presented of upwards of three millions of people receiving rations from government. On that memorable occasion, all that the highest talent for organization and the warmest zeal and the most ample expenditure and the most unsparing labour could accomplish was done, for the lives of a whole nation were at stake. Still the most active administrators of relief all agree in acknowledging the failure of their best exertions, and at the same time declare that with all their experience of the past, they could not if again placed in similar circumstances hope for any greater success. There was no want of food, for it was imported in immense quantities; there was no want of money to purchase it, for public and private charity supplied funds with unsparing hands. The overwhelming difficulty of that time was the impossibility of distributing relief, arising mainly from the non-existence of provision dealers throughout the most distressed districts.* Thus the most elaborate organization of civilized man was unable to effect even for a short time, and on one great emergency, what the spontaneous action of society daily and perfectly accomplishes.

§ 9. It is not however sufficient that means should exist of bringing to the same place at the same time a number of persons desirous to exchange. Means must also be found of placing them in communication. In the general resort of trade, individual buyers and sellers must in fact notify to each other their willingness to transact business. Various methods have been devised for effecting this important purpose. The most ancient perhaps, as it is the

---

* *Transactions of Central Relief Committee of the Society of Friends*, pp. 20, 99, 103.

most obvious, finds its representative in our modern street cries. In our thoroughfares and in the less wealthy part of our towns, the hawkers, the modern types of a remote and inferior development, loudly proclaim the nature and the merits of their wares. Three centuries ago this mode of business was even in London universally prevalent. But this system of oral advertising was both troublesome and intermittent; and consequently men sought to announce by some permanent form both the fact that the house was one at which sales were made, and the character of the wares which it contained. This information was given to an illiterate people by means of sign boards. It was one of the characteristics of English towns in the last century, as it still is of many continental towns, that their streets presented rows of sign boards of various kinds, but all more or less indicative of the inhabitants' occupation. Even still the "prisci vestigia ruris" linger among us; and in the barber's pole and the sign of an occasional shop or inn we retain the mode of advertising that belonged to the era antecedent to the age of newspapers and of national education.

When the name of the inhabitant and the style of his occupation superseded the old sign board, the pressure of competition continually forced sellers to contrive other means for attracting public attention. The handsome window, the tempting display of commodities, the open door that invites entrance, all these and a thousand similar artifices are now almost universal. In those occupations in which custom prohibits direct modes of advertising, ingenious contrivances for becoming known are continually adopted. All these plans, whatever they may be, are designed to attract by some means public attention. The occupation of the advertiser is easily made known; and his object is to be the subject of constant conversation. Notoriety, like abuse, if applied in sufficient quantities, is sure to adhere. It is to their observance of the power of advertising that not a few inferior men in almost every walk of life have owed their success: and it is to their neglect of it that many men of real ability and skill have pined in undeserved neglect.

But although many modes of attracting public attention are now adopted according to the usages of different occupations, the term advertisement is usually limited to a printed announcement. That remarkable system of advertising with which we are so familiar is like many other social phenomena of very recent date. The earliest known advertisement bears date about five weeks after the tragic death of the First Charles, and relates to some stolen horses in Suffolk.* Another advertisement that for some time claimed the honour of priority was the announcement of an "heroic poem" in 1652. It was long however before this system attained anything like its present proportions. Most men of business were then too poor to advertise. The system was still a novelty, and its advantages were not appreciated. The reading public too was very limited. Only a comparatively small part of the public could read. The numbers of these whom the newspaper reached were still less. Nor was it of much advantage for a London tradesman to notify to rural districts the excellence of his commodities, when there was no convenient means of delivering to such customers his goods. The shortsighted jealousy of the Government also seriously retarded their means of communication. Every newspaper paid stamp duty: every advertisement paid a separate and heavy tax: the paper on which it was printed was taxed. But notwithstanding these difficulties, as the wealth and intelligence of the population increased, the newspapers found more readers and brought to them more advertisements. The advertise-

---

* Andrew's *History of British Journalism*, vol. i. p. 49.

ments in an ordinary number of *The Times* may now exceed 2,500.* In addition to the regular advertisements in the newspapers and other recognized channels of communication, it is said that two millions stirling are now annually spent in England in circulars, handbills, placards, and other modes of "extra advertising." The sums paid by some mercantile houses for advertising are very large. The annual expenditure of one well-known advertiser is said to amount to £40,000. Three London firms are mentioned as expending for this purpose £10,000 each.† Advertising indeed seems to be rapidly attaining the position of a distinct art. Its importance seems now to be tolerably appreciated. But the process is both costly and difficult. The figures I have quoted sufficiently attest the cost. The difficulty may readily be understood. No small judgment and experience are required to determine in each particular case the best form of advertisement and the best advertising medium. Much expense may be absolutely wasted in advertisements which fail to reach the bulk of the classes that are likely to become purchasers.

From this costly necessity of advertising an unforeseen consequence follows: the wealthy producer obtains a great additional advantage over his competitor who is less able to advertise; and thus production upon a large scale is again more profitable than production on a small scale. A remarkable illustration of this tendency is found in the case of books. It costs as much to advertise a sixpenny pamphlet as it does to advertise the most expensive work of reference. Hence publications of the former class are seldom if ever remunerative. The expense of advertising the eighth edition of the *Encyclopædia Britannica* is said to

---

* Ib., vol. ii. p. 338.
† Ib.

have amounted to £3,000. No smaller work could have borne such a charge. So too that "keeping up appearances" which forms the advertisement of a professional man falls heavily upon the less successful man: while his more fortunate competitor both gets business because he appears prominently before the public, and continues to appear before the public by the business that he obtains.

§ 10. Another arrangement for facilitating exchange, and upon which the success of advertising largely depends, consists in the system of public communication. This depends in the first place upon the state of the roads and the means of conveyance; and next upon the quality of the postal arrangements. Of the former I shall presently have occasion to speak. The latter seems to require separate consideration. Yet this peculiar notice is not necessitated by any peculiarity of the subject itself. It arises from external circumstances. The carrying of mails is merely a branch of the ordinary trade of carrying. It arises spontaneously in the ordinary growth of a community. But since, in almost every country, the government has claimed a monopoly of this business, men generally regard it as one of the ordinary functions of government, and forget that the post-office is but of very modern date, and that neither its origin nor its improvement are due to official ingenuity or official zeal. In some of the more remote localities of Australia, where the settlers reside at great distances from each other and from any post-office, they combine to employ a horseman who is constantly engaged in carrying the post from the nearest post-office to their respective stations. There is no reason to doubt that this effort to supply an acknowledged want would, if circumstances required, be made upon a larger scale: and that if government were now to discontinue the business of carrying mails, or even to abandon its monopoly, private enterprise would speedily supply its place. The principal advantage, apart from any supposed political expediency, that arises from the interference of government, seems to be the extension to the poorer districts, at an earlier period than if unaided they could have obtained them, of the benefits of cheap communication. If the service, as a whole, pay its expense, post-offices and telegraph stations may be maintained where a private company would not keep them; and thus without any perceptible burden to the richer localities, the more remote and less prosperous districts are brought within the range of a powerful agent of civilization. The importance of this agency and its recent expansion may be inferred from the following statement. In 1837 the number of documents of all kinds including public statutes, and newspapers that passed through the post-office in the United Kingdom amounted according to the most careful estimates to nearly $126\frac{1}{2}$ millions. In 1860 the number of letters actually delivered was 564 millions; of newspapers nearly 71 millions, and of book packets nearly 12 millions. About two and a-half millions of undelivered papers and letters must be added to this amount.[*] After every deduction has been made for all the other purposes of correspondence there still remains in this vast mass an immense residue which we must fairly consider as directly tending to facilitate exchange.

§ 11. When a market has been found and the terms of the exchange settled, another consideration still remains. Provision must be made for delivering at that market the subject of the exchange. The means for effecting this purpose that are generated in the progress of society are the deve-

---

[*] Levi's *Annals of British Legislation*, vol. ii. p. 502.

lopment of the carrying business and its subsidiary branches, and the construction and improvement of roads, ships, canals, and other modes of communication. In the earlier stages of social development the producers sought their own market and conveyed their own goods. They combined the various functions of producer, merchant, and carrier. But as the operations of producing commodities and of seeking a market for them when produced are not only difficult but dissimilar, a distinct class of men by degrees devoted themselves so the latter task, and thus the separation of functiont began. Accordingly the mercantile class, a body of men intervening between the producer and the consumer, was gradually formed. This class undertook the double function of finding a market and of reaching it. They both conducted the sale and transported the goods. Such has been from time immemorial the manner in which traffic has been carried on in the East. There the merchants unite into caravans, under a leader of their own selection, under a rigorous discipline, to pass safely through the formidable desert and the still more formidable marauders. It is in this way that the Jews and the Armenians, and especially the indefatigable and skilful merchants of Bokhara, carry on the great and increasing trade between the Russian dominions and the tribes of Central Asia; and that European goods find their way to the frontiers of China and Thibet. Even in our own society in those parts which are still but imperfectly developed, we may recognize the same type. The pedlers of the more remote districts of Scotland and Ireland are the analogues of the Eastern or the mediæval merchant. We may find a still more complete illustration of the rudimentary condition of the distributive class in the street-folk of our metropolis. Such are the groundsel-sellers or the chick-weed gatherers, who themselves procure their commodities, search the streets for a customer, and convey

2 B

their humble wares to his house.* But as society advances, the occupation of the carrier is separated from that of the merchant ; and becomes sufficiently important to require the undivided attention of a peculiar class of men. In course of time as the business of carrying grows, it in its turn, like the business of the merchant, is again subdivided; and in addition to its own leading divisions gives rise to many subsidiary employments. There are carriers by land and carriers by water : there are innkeepers who provide temporary accommodation that carriers require. There are agents that collect goods and deliver them to the carrier, or in turn receive them from him. And there are the numerous classes of tradesmen and porters and assistants that the construction of the carriages and the other wants of a great carrying establishment involve.

No changes have been more remarkable or more widely influential than those in the business of carrying. But these changes are most conspicuous and most obviously dependent on social progress in the case of land carriage. The sea requires no such preparation of its fleets as that which the surface of the earth demands. Those great artificial means of inland navigation which half a century ago formed so remarkable a phenomenon in British industry, although not superseded by recent inventions, are never likely to be extended. The conditions of our improved navigation, increased knowledge of physical science, the use of more powerful means of propulsion, greater skill in shipbuilding and in navigation, and it may be added in the management of passengers, and of crews, are merely different applications of the same principles as those which govern land conveyance. I shall therefore even at the disadvantage of repeating a thrice told tale, briefly indicate, so far at least

---

* Mayhew's *London Labour*, vol. i.

as land carriage is concerned, the various stages of this progress.

The first instrument of conveyance is the human back. Such is the mode of transport in use not only among all barbarous tribes, but among those inferior types in our own population that most nearly approach the verge of civilization. No small advance has been made when for human labour, the labour of some lower animals, the horse or the ass, the camel, or the reindeer, has been substituted. In such a case, if the natural surface of the country be at all favourable, a cart of some kind quickly follows. "It is calculated," says Dr. Lardner,* "that a horse of average force working for eight or ten hours a day cannot transport on his back more than two cwt., and that he can carry this only at the rate of twenty-five miles a day over an average level country. The same horse working in a two-wheel cart will carry through the same distance per day twenty cwt. exclusive of the weight of the cart. By this simple expedient therefore the art of transport was improved in the ratio of one to ten : in other words the transport which was before effected at the cost of £10 was with this expedient reduced to the cost of £1." It is not therefore surprising that, in all countries in which beasts of burden are found, the use of some description of vehicle has also been adopted. But there are few countries which have advanced much beyond this primitive stage. Any further advance depends upon the presence of roads, and roads are surprisingly few. It is said that at the present day in two-sevenths of the inhabited part of the globe, there are no roads at all. In a great part of those countries which in this estimate are assumed to possess them, the roads are merely nominal. There are few of these countries more advanced at the present day

---

* *Railway Economy*, p. 231.

at least in their less frequented districts than Lancashire and the South of Scotland were eighty years ago. There is ample evidence as to the means of communication which these great seats of industry at that time enjoyed. It may not be uninteresting to those whose roads are made to hear once more how much they owe to those means of communication that they use with so little thought of their utility; and at the same time it may afford some small consolation to the settlers in a new country that their sufferings are not greater than those which their grandfathers endured in the good old times of George the Third.

"It is not easy," says Mr. McCulloch,* "for those accustomed to travel along the smooth and level roads by which every part of Scotland is now intersected to form any accurate idea of the difficulties the traveller had to encounter in that country a century ago. Roads were then hardly formed, and in summer not unfrequently consisted of the bottoms of rivulets. Down to the middle of last century, most part of the goods conveyed from place to place in Scotland, at least where the distances were not very great, were carried not by carts or waggons but on horse-back. Oatmeal, coals, turf, and even straw and hay were conveyed in this way. At that period and for long previously single-horse traffickers (cadgers) regularly plied between different places, supplying the inhabitants with such articles as were then most in demand, as salt, fish, poultry, eggs, earthenware, &c.; these were usually conveyed in sacks or baskets suspended one on each side of the horse. But in carrying goods between distant places it was necessary to employ a cart, as all that a horse could carry on his back was not sufficient to defray the cost of a long journey. The time that the carriers, for such was the

---

* *British Empire*, vol. ii. p. 58.

name given to those that used carts, usually required to perform their journeys seems now almost incredible. The common carrier from Selkirk to Edinburgh thirty-eight miles distant required a fortnight for his journey between the two places going and returning. The road originally was among the most perilous in the whole country: a considerable extent of it lay in the bottom of that district called Gala Water, from the name of the principal stream, the channel of the water being when not flooded, the track chosen as the most level and easiest to travel in." Nor were the means of locomotion in the southern parts of the island in a much better state. It will suffice out of many instances to repeat a well-known passage from Arthur Young's travels in which he describes no mere mountain track, but the actual turnpike road between Preston and Wigan as he found it in the year 1770. "I know not," he says, "in the whole range of language terms sufficiently expressive to describe this infernal road. To look over a map, and perceive that it is a principal one not only to some towns but even whole counties, one would naturally conclude it to be at least decent; but let me most seriously caution all travellers who may accidently purpose to travel this terrible county to avoid it as they would the devil, for a thousand to one but they break their necks or their limbs by overthrows or breakings-down. They will here meet with ruts that I actually measured, four feet deep and floating with mud only from a wet summer; what therefore must it be after a winter? The only mending it receives in places is the tumbling in some loose stones which serve no other purpose but jolting a carriage in the most intolerable manner. These are not merely opinions but facts, for I actually passed three carts broken down in these eighteen miles of execrable memory."*

---

* Porter's *Progress of the Nation*, p. 296.

§ 12. The vehicles and the horses of our ancestors were not out of keeping with their roads. There is an obvious connection between the state of the roads and the state of the coaching in any country. Even without regard to the injury that bad roads cause both to horse and to carriage, a well-appointed vehicle has its power immensely increased by being placed on a well-made road. The only means by which the expense of making a good road can be quickly repaid is by a considerable increase of speed; and that increase a heavy carriage or bad horses can never attain. We therefore find that improvements in each branch quickly follow each other. The sequence however varies in different circumstances. When some kind of road is in existence, as in most long settled countries, the improvement begins by repairing the road; and then the goodness of the road brings after it more convenient conveyances and speedier horses. Where on the other hand no roads exist, but where there is a strong demand for the means of intercourse, as in many parts of America and Australia, the best and speediest conveyances that the case will admit are contrived; and the roads are gradually made as time and means allow. In England, which may be fairly regarded as the most favourable instance of improved communication in an old country, pack horses were for a long time the sole means of transport. The next step consisted in the use of long wagons for passengers and goods. These conveyances began about the time of Queen Elizabeth to travel to London from Canterbury and other large towns. Then in the reign of Charles II. stage coaches were introduced; and continued for upwards of a century to afford the most rapid means of locomotion then known. The original character of these vehicles may be best understood from marking the improvements which in their day caused unbounded delight and admiration. "We have before us," writes Mr. Knight,* "a very curious bill of the Alton and Farnham machine, dated 1750, which is headed with an engraving furnishing the best representation of the coach of a century ago that we have seen. This clumsy vehicle carries no passengers on the roof; but it has a large basket, literally a basket, swung behind for half price passengers. The coachman has four horses in hand, and a postilion rides a pair of leaders. This is truly a magnificent equipage, and it accomplished its journey in a marvellously short time, starting at six in the morning and arriving duly the same night. This journey of forty-seven miles in one day was a feat, and well might the vehicle which accomplished it be dignified by the name of "machine." The name became common; and hence stage-coach horses were called "machiners."

All the improvements in locomotion in England followed some attempt to improve the roads. The long wagon of the early days of Elizabeth followed the attempt in the reign of Philip and Mary to enforce the obligations of the parishes to repair the highways. The flying coaches of the time of Charles II. were the consequence of the Turnpike Acts. It was not until turnpike roads became extended that the next great improvement of mail coaches was sanctioned, shortly after his triumph over the Coalition, by the younger Pitt. And it was not until the days of Macadam that the improvement in the ordinary stage coaches kept up with that effected by Mr. Palmer in the mail. From recent changes the roads are now chiefly useful in connecting lines of railways; but it will shew how great has been the progress of England in this respect when we remember that thirty years ago in England and Wales half a million of acres had been turned into metalled roads, and that the

---

* *Curiosities of Communication*, p. 15.

usual rate of travelling on these roads was ten miles an hour.*

§ 13. Since the advantages of roads are so great and so conspicuous, we may naturally feel surprise that in so large a portion of the world their use is unknown; and that in those countries where they have been employed, their improvement has been so slow. The reason seems to be that the construction of roads requires considerable wealth, and considerable powers of co-operation. These conditions, at least during past ages and in most countries, are seldom found. Except in the case of the Romans, whose roads were made solely for military purposes and with means forcibly acquired from conquered countries, and cannot therefore be regarded as instruments designed to facilitate exchange, we do not find any attempt at road-making among the nations of antiquity. The utmost efforts in this direction of the populous nations of the East seem to have been the use of pioneers to prepare the way for some great chief or some advancing army. These preparations consisted merely in the removal of the natural obstacles, and did not involve any of that elaborate construction which is so familiar to our engineers. For the slow advance of road-making, we may at least in England trace the operation of other causes. The traffic in early days was but small; and although this smallness was to some extent the consequence of difficult communication, the existing channels were upon the whole sufficient for its moderate demands. The state of the law too presented a serious impediment to the improvement of roads. Each parish was bound to maintain the roads that ran through it. The parishes might have been able, and perhaps willing, to repair at least sufficiently for

---

* Porter's *Progress of the Nation*, p. 29.

their own purposes the damage done to their roads by their own traffic. But the scanty inhabitants of remote rural districts could hardly be expected to repair a road worn by the traffic in which they were little interested between two populous towns. Accordingly in 1663 the first Turnpike Act was passed not without opposition, and was carried into effect not without bloodshed. So true is it that "unjust and absurd taxation to which men are accustomed is often borne far more willingly than the most reasonable impost which is new."* Even when the legal obstacle was thus removed, upwards of a century elapsed without any very marked improvement. At length the effects of the great industrial impulse that followed the Peace of Paris became apparent, and men found that they could no longer go on under the old system. The extension of our industry created a demand for improved communication, and improved communication quickly assisted to develope industry. Thus the demand for good roads and speedy coaches grew, until as usually happens it called forth the abilities of men competent to supply the want. The improvements in road-making of Macadam and of Telford were seconded by improvements in coach-building, and in the quality of the horses used; until both roads and horses were superseded by a smoother way and a more potent agent.

§ 14. I do not propose to discuss at large the history or the advantages of steam. Its wonderful results are patent to all. It will be sufficient to observe that the application of steam to locomotion formed a condition precedent to other improvements to which I have already referred; and itself too depended upon the combination of several antecedents. Without the facilities that railways afforded, postal

---

* Macaulay's *History of England*, vol. i. p. 374.

reform would have been either wholly impossible or miserably imperfect. Without railways, the communication between buyers and sellers, manufacturers or wholesale dealers and retailers, must still have been conducted by commission agents. But the motive power of steam had been long known, and the miners of the north of England for more than two centuries before the first Railway Act was passed had used a primitive railway. Before these ideas could be combined, before the steam-engine could be driven on the line of rails, other conditions were requisite and were gradually fulfilled. During the latter half of the last century, notwithstanding the efforts made to keep pace with the growing demand, and doubtless aided by those very efforts, the traffic had outgrown the old method of communication; and increasing demands were made not only for increased power of production but for increased power of exchange. New means of smelting iron, the discovery of new beds of coal, a growing conviction of the importance of roads, an abundance of capital, and increased knowledge of physical science, and increased skill in its application, all these were necessary phenomena in a state of society in which the germ of thought destined to grow into the modern system of railways should continue its development. Already 9000 miles of railway representing a capital exceeding three hundred and fifty-five millions of pounds are opened in the British Isles; and nearly three times that number of miles have been constructed in North America. In the former case at least, much of that immense outlay has been wasted in litigation and in other ways not fairly chargeable to railway purposes. Yet although, if the work were to be done again, it could be done with greater efficiency and for less money, the saving that has been effected is very great. Less capital is in transit, less capital is kept idle in superfluous stock. Less time is spent in travelling. New markets are opened;

and in old markets prices are equalized. It has been found that on the opening of a line for passenger traffic between two places, the amount of travelling between the extremities of the line has been generally quadrupled.* On the short line between Liverpool and Manchester the annual saving to the public, irrespective of time and of the extra gains of the proprietors, is estimated at about a quarter of a million of money. As regards the increased facility of communication and the consequent saving to the traveller of time and money, a curious example is given by Mr. Disraeli.† Mr. Robert Weale, an Assistant Poor Law Commissioner, travelled during twelve years in the performance of his official duties 99,607 miles: of this distance 69,000 miles were by coach at an average cost of 1s. 6¾d., and 30,000 were by rail at an average cost of 3¼d. If the whole distance had been done by rail it would have taken one year, thirty weeks and six days; if by coach, four years, thirty-nine weeks, and one day. Thus Mr. Weale would have saved three and a-half years of his life, and the country £5,390 in his travelling expenses alone irrespective of his salary during the time he was engaged in travelling.

§ 15. There is, as Adam Smith has remarked, a striking analogy between roads and money. This analogy arises from their common purpose as machinery of exchange, and might in like manner be traced between the other arrangements for the same object which men have designedly adopted. They are merely instruments contrived to facilitate exchange. Although they are conditions precedent to extensive production, they do not of themselves produce anything. They are, in the language of Mr. Mill respecting money,

---

* Porter's *Progress of the Nation*, p. 332.
† *Life of Lord George Bentinck*, p. 409.

"merely machines for doing quickly and commodiously what would be done though less quickly and commodiously without them." They are most potent instruments, and in their full development produce very remarkable results; but they are expensive instruments, and they produce those results indirectly by stimulating other sufficient agents. Like all other instruments, they should both in quantity and quality be sufficient and no more than sufficient to effect the required object. Their practical utility therefore in any case depends not upon their absolute magnitude, but upon their magnitude as compared with the number of exchanges likely to be effected. There are consequently two faults to which the use of both roads and money is liable. They may be insufficient for their object, or they may be more than sufficient. In the absence of interference, the latter evil is not likely to arise. Where roads are made or money provided under the ordinary conditions of industry, a greater amount will never be permanently supplied than the exigencies of society require. A deficiency might in some cases be experienced, since they require a greater capital and a greater amount of co-operation than infant societies usually possess. In such cases the interference of government, if judiciously applied, may accelerate and assist the ordinary processes of natural development.

Unfortunately the judgment shown in this interference is seldom very sound. Money has been issued in such quantities as to produce a ruinous depreciation, and of a quality very different from that which it is represented to possess. Roads are often made either in a direction different from that in which they are really wanted, or on a scale unsuited to the requirements of the traffic. A road of the latter description is analogous to an excessive issue of money, except that the superfluity is not so easily disposed of. If capital, whether it belong to one person or to a multitude of persons, be sunk in procuring an instrument, too good for the work to be done, that capital is wasted : and the proprietor has so much less to expend on other purposes. It is immaterial whether the instrument be a ton of gold, where a quarter of a ton would suffice ; or a railway, where the traffic only required a bush-track; or a steam-engine, where the work could be done with a hand-mill. But as no issue of spurious money will prevent men from returning to the old and well-tried medium of exchange, so no deviation from the shortest line will permanently exclude traffic from its natural channels. "The experiences of America," it has been observed,* " in relation to public communications proves beyond perhaps the experiences of any other part of the world the fact that the speediest, cheapest, and most convenient, routes from one great source of business to another will in the end be adopted. There is hardly a state in the American Union which does not furnish more or less examples of the short-sightedness of legislatures in providing for the wants of the future. Railroads have been projected and made, time and again, to meet the wants of thousands. Before they were ten years in operation, millions required railroad facilities. Local interests, and local ignorance have almost everywhere caused roads to wind round to one out of the way place, or to take an unnatural route to another. But the waves of population directed by a higher sagacity moved in the direction of the rich lands and the fertile country, and left the petty roads to be but a reproach to their concoctors or a burden to the people. As a general rule, a really great work, something that American progress justified, no matter how it might have been underrated in the legislature, has been certain to prevail in the end."

---

* Hogan's *Prize Essay on Canada*, p. 54.

CHAPTER XXI.

OF THE INDUSTRIAL EVOLUTION OF SOCIETY.

§ 1. In every organism there have been found to exist certain well-defined and fundamental laws of growth. Every creature that has life commences as a small aggregation of cells. Each of these cells attracts, assimilates, and organizes the nutrient particles of the surrounding blood. It thus gradually increases in size, and at the same time it prepares a portion of this nutriment to serve as the germ of new cells with endowments similar to its own. Although a constant decomposition of the structure is in progress, and ultimately prevails, yet for some time the process of nutrition considerably overpowers its rival; and under its influence the structure increases in bulk, until in its maturity it exceeds often in an enormous degree its original dimensions.

In its rudimentary condition the structure of the organism is so simple as hardly to deserve that name. Its development however brings with it, or more accurately consists in, an increased complexity. Every germ consists at first of a substance which both in texture and chemical composition is perfectly uniform. A difference soon presents itself between two parts of this substance; and each of these parts present in their turn similar differences. When the process of differentiation once sets in, its advance is rapid. Each differentiated part becomes itself the subject of new differentiations; and so the development proceeds in geometrical progression, until out of the simple, vague, and homogeneous germ, the adult plant or animal with all its complex combinations of distinct tissues and organs is evolved.

The earlier stages of development exhibit a uniformity not only of structure, but of function. Both in its form and in its action, each part resembles every other. Each is uniform, and each is independent. But the course of development is marked by an opposite set of conditions. The homogeneous structure gradually becomes heterogeneous; and the uniformity of function gives way to variety. The division of employments is established between the several parts of the organism. Their separate existence is merged in the larger collective life; and they become the component parts of an organized whole. Thus a mutual interdependence of parts is established; and while in the undeveloped germ each portion exercises its own independent powers, the activity and the life of the several parts of the adult creature are rendered possible by the activity and the life of the remaining parts.

The degree in which these characteristics exist in any organism determines the relative position of the creature in the scale of existence. "The more imperfect a being is," says Goethe, "the more do its individual parts resemble each other; and the more do these parts resemble the whole. The more perfect a being, the more dissimilar are the parts." It is therefore in the ascending order of their complexity that naturalists have classified the vegetable and the animal kingdoms. Nor is the reason of this principle obscure. The multiplicity and the variety of the special organs are the condition of a richer life and a completer growth to the animal. They are, as it has been well observed,* "So many different means whereby the individual may place himself in relation with the external world, may receive the most varied impressions from it, and so to speak may taste it in all its forms, and may act upon it in turn. What an immense distance between the life of a polype, which is only a digestive tube, and that of the superior animals; above all, of man, endowed with so many exquisite senses, for whom the world of nature, as well as the world of ideas, is open on all sides, awakening and drawing forth in a thousand various ways all the living forces wherewith God has endowed him."

§ 2. The same phenomena which thus characterize the evolution of an individual may be observed in the evolution of society. In both cases the evolution consists in, or at least is invariably attended by, an increase of bulk, a greater complexity of structure, and a consequent interdependence of parts. The social germ, like the individual germ, has a simple and uniform origin. It too attracts, assimilates, and organizes other like bodies; and thereby grows in bulk. It too presents with every increase in bulk an increase of complexity; and continually substitutes for a single indefinite and homogeneous structure a series of well-defined and mutually dependent organs. Each of these organs performs its special function; each function acquires its appropriate organ. Each organ is related both to the whole organism, and to every other organ; and depends for its efficiency upon their co-operation. Society therefore, like the individual organism, tends to become more complex, and its parts consequently become more closely interdependent. The more complex is the individual organism, the greater is its activity; and the keener is its sense of enjoyment. So too

---

* Guyot's *Earth and Man*, p. 97.

in the social state, it is with the advance of civilization, and in proportion to it, that these characteristics of development are observed. If we count the members of a barbarous tribe by hundreds or by thousands, we count the numbers of the great civilized nations by millions. In the complexity of its social structure, and in the definite character of its several organs, no less than in its actual bulk, the English nation exceeds an aboriginal tribe at least as much as one of the higher mammals exceeds a zoophyte. In some of the lower forms of life, so loose is the relation of parts that when the creature is cut into two each portion still preserves its vitality; while in the more highly organized animals the loss even of a small portion of the structure causes always constitutional disturbance and often death. In the same manner the union of the small and undifferentiated groups of families that form an aboriginal tribe is inconstant and easily severed. Even where a rudimentary organization exists, its extent is for a long time limited; and the differentiations are imperfect and easily effaced. But in those countries which are confessedly the foremost in civilization, the effects of a benefit or of an injury done to one class are rapidly felt throughout the whole community. The insolvency of the capitalist throws the labourer out of employment. The prosperity of the labourer brings with it an increased demand for the use of capital and increased profits. When times are good, rent is well paid. When rents are well paid, there is a fresh stimulus given to profits and to wages The bankruptcy of even one great firm sometimes, as we know, involves in its fall numerous distant, and in appearance even unconnected, establishments.

§ 3. Life implies constant action; and action involves waste of tissue. Every organism therefore requires a continual supply of nutrient material sufficient to replace its incessant disintegration. But the increase of bulk in an organism is effected by the formation of new tissues; new tissue, as well as that which replaces preceding tissues, in like manner requires nutriment; and this nutriment is not supplied until the wants of the existing tissues have been satisfied. Consequently new tissue can be formed only to the extent of the surplus nutriment after the reparation of the ordinary waste of the system. The growth therefore of any organism within the limits which its original constitution imposes, depends upon the supply of appropriate nutriment that it can obtain. If that supply be merely equal to its consumption, it will continue to live, but will not grow. If the supply be in excess of the consumption, its growth will be proportionate to that excess. If the supply fall short of the demand, the organism will dwindle, and will ultimately die. The same principle which controls the growth of the whole organism applies equally to its parts. If the functional activity of any organ be stimulated beyond that of any other organ, the size of the one will increase, and that of the other will decrease. The exercise attracts to the active organ a larger quantity of blood and consequently of nutriment; and when the quantity of blood in the whole organism remains unchanged, the required amount is supplied at the expense of the less active organs. Thus the one is enlarged on account of the additional nutriment it receives, and in proportion to it. The others for the same reason and to the same extent are sacrificed. If however the former organ be worked so hard that all its channels of supply cannot furnish blood sufficient for its repair, both the organ dwindles and the whole body is impaired.

If from a single organism we extend our view to a collection of organisms, we shall find the same principle in operation. The numerical increase of such a collection obviously implies and is limited by the replacement of the original number. For this purpose there must be a sufficiency of the necessaries of life for that number. If there be more than such a sufficiency, the number may to that extent increase. If there be less than such a sufficiency, the number must be proportionately reduced. These necessaries of life include air, light, heat, and sustenance. The supply of these agents must be unfailing and regular; their quantity must be abundant; and their quality must be pure. These terms are obviously relative. There may be a sufficiency of means to maintain life, or even to admit of growth; but the life may be weak, and the growth slow. But when the supply of these agents at all times is of the best kind, and as ample as the organs can conveniently use, the conditions of development are favourable. The presence therefore of all these conditions marks the extreme limit to which each species can increase. The numbers of every kind of organism may advance up to the point at which each individual has a sufficiency of these primary agents, but not farther. Yet although an impassible limit is thus marked, it is seldom indeed that it is attained. Great as is the fecundity of both plants and animals, there are positive and very stringent checks in operation to control it.

Once at least, if not oftener, during every year, in their natural state, almost every plant produces seed; and almost every pair of the lower animals have offspring. There is therefore a constant tendency both in plants and animals to increase in a geometrical progression. If this tendency were unchecked, the earth would speedily be covered with the descendants of a single pair. Apart from any merely arithmetical calculation, it is an ascertained fact that in favourable circumstances many plants and animals have actually in a short space of time spread in astonishing numbers over very large tracts of country. Nor is there any material difference in this respect between the slow breeding organisms, and those whose produce is counted by myriads. The question is merely one of time. The slow breeder would require a few additional years to fill, in favourable circumstances, the largest districts. In their case the geometrical rate of increase still continues, but the number of the rate is comparatively small. That there is still standing room on the earth must therefore be due to the action of some repressive measures.

These measures are found chiefly in the mutual limitations produced by the several species. Each species serves as the food of that above it in the scale of creation. Even between individuals of the same species, either in the ordinary distribution of food, or still more in seasons of scarcity, the feebler are as it were pushed aside, and trampled down by the stronger. Circumstances that favour or discourage the increase of one class of animals will materially modify the condition of a different but related class. Thus the destruction of small birds in France has by removing the check from the multiplication of insects caused such serious injury to the vines of that country that legislative interference has been required. In like manner in Paraguay the absence of insectivorous birds permits the increase of certain flies that are destructive to young animals; and so limits the number of cattle in that country, and thereby affects the character of its grasses. It has been remarked that the presence of cats in an adjacent village by limiting the number of mice, as the mice limit the numbers of bees, whose agency is essential to the fertilization of certain plants, influences materially the flora of a district. Concurrently with this great check there are others in steady though less frequent operation. Immense numbers of creatures are known to perish in seasons of extreme inclemency. The want of sustenance, and the want of shelter are in the lower departments of nature not the least influential checks upon excessive increase. Hence it is easy to understand that the average number of any animal or plant depends only indirectly upon the number of its eggs and seeds. But I do not now propose to consider this most wonderful and most complex system of checks and counterchecks. That which is material for the present purpose is that every organism is striving to increase in a geometrical ratio : that either during the whole of its life or at some particular part of it, either during each generation or at brief intervals, it is subject to great destruction ; that by this destruction its numbers are kept, with but temporary variations, up to a certain average ; and that, unless there be some change in its conditions, this average is not permanently exceeded. If however the conditions be changed, if the pressure of any check be even for a little lightened, if the destruction be ever so slightly reduced, the number of the species will increase with incredible rapidity and to an incredible extent.*

§ 4. To these laws of vegetal and animal life man presents no exception. He too, like other organized beings, is constantly striving to increase at the geometrical ratio. He too, although a slow breeder, has given ample proof that if this rate of increase were unchecked he would soon replenish the whole earth. He too cannot increase beyond that barrier which nature has fixed for all created things : and he too, whatever may be his power, never does in fact, while his command over the primary conditions of existence is unchanged, exceed an average which, unless in rare and exceptional circumstances, admits of but little variation.

Although these general points of resemblance exist,

---

* See Mr. Darwin's *Origin of Species*, ch. iii.

there is also a material distinction, arising from the difference in their natures, between the increase of man and that of other animals. The rate of increase in both cases is checked, but the checks are very different. In the lower organisms the check is positive. It consists in the death from various causes of living beings that have actually come into existence. Excepting perhaps a few of the larger mammals, each species is the food of that above it; and the destroyer in his turn affords nutriment to a more highly organized destroyer. The animal kingdom finds its type in the warrior priest of the Arician grove, "The priest who slew the slayer and shall himself be slain." But there is no species of which man, the crown of animal life, is the habitual prey; or which can materially reduce his numbers. Man too has the power of artificially increasing his supply of food; and of mitigating, if he cannot wholly avoid, the severity of those climatic influences which so powerfully reduce the numbers of other animals. In a civilized community therefore, the positive check is to a great degree at least, if not wholly, removed; and so far as regards its operation the human race would inevitably in a few generations exclusively occupy the whole surface of the earth. But in the same compensating power of mind which thus places so feeble an animal as man above the reach of elemental forces or of natural enemies, there arises a new and most efficient check. While in the lower organisms the excess of their numbers dies, such an excess in the human species, or at least in the more advanced portion of it, is never born. Man seldom *will* do all that he *can* do. He has indeed the power which all other animals possess of annually producing offspring. But he does not act, as they act, under the impulse of blind instinct. He knows, and he forbears. If he think that by a temporary abstinence he can secure, more completely than he otherwise could, for himself and those dependent on him a sufficiency of the necessaries of life and of what in his estimation are its comforts, he will refrain from marriage until his object be accomplished. Besides, his standard both of sufficiency and of comfort constantly tends to rise. Men desire that their children shall occupy no worse position in life than they themselves possess. Until therefore a man can see a reasonable prospect of providing for his future family not merely in such a manner as to enable them to support existence, but in a manner suitable to the degree of comfort in which they were born, he will not incur the responsibilities of marriage. Thus there arises a preventive check of the most powerful kind. This check springs from the very nature of man; and its potency increases as the distinctive characteristics of man increase. It is in advanced communities that the operation of this check is most effectual. In savage life and in the lower forms of civilization, its influence is comparatively slight. In these cases the mortality is very great. The nearer that man approaches the state of the lower animals, the greater becomes the influence upon him of the mere animal or positive check; and the feebler is the operation of that intelligent forbearance and self-control which is the chief glory of our race.

§ 5. While the increase of population is thus limited, men are at the same time continually endeavouring to lighten the pressure of the check, by enabling an ever-increasing number of persons to comply with the requirements that prudence prescribes. If wealth bring with it population, population in its turn brings with it wealth. The most favourable circumstances for the exercise of all the industrial aids, but especially of co-operation and exchange, are found amidst a large population. It is only where a large population is assembled that combination of labour, on such a scale as the execution of great works of construction require, is possible ; or that the ordinary combination for the usual industrial occupations can be general. It is only in a large population that a perfect organization of labour can be effected ; or that the facilities for exchange can be complete. It is only in such circumstances as these that large accumulations of capital are made ; or that inventions can be executed or rewarded. From the wealth and the power thus obtained a large population becomes capable of exercising a more complete control over nature, and of converting to its use superior natural agents, than before. A large population therefore, if it be duly organized, ought to be more wealthy, and consequently to have a greater capacity for increase, than a small one.

This rule includes an important proviso. The large population must be duly organized. The mere increase of a homogeneous mass does not fulfil the required conditions. It is a general principle in the increase of organisms that the same space of ground can support a much larger amount of life if the species be different, than it can if there be but a single species. The greatest amount of life in a given space is obtainable by the greatest diversity of structure and of habits. If all the inhabitants of a district required the same objects, there would be an intense competition for these objects while the other products remained unused. Experiments have proved * that if a plot of ground be sown with one species of grass, and a similar plot be sown with several distinct genera of grasses, a greater number of plants and a greater weight of dry herbage can be raised by the latter means than by the former. The same rule has been found to hold good when on equal spaces of ground, first one variety, and then several mixed varieties, of wheat have been sown. So also is it with the lower animals. In like manner

---

* Darwin's *Origin of Species*, p. 1¦3.

the largest and most flourishing populations are found where there is the greatest diversity of employment, and the complaint of over-population is seldom heard except when the occupations are nearly uniform. This consideration explains the greater impulse given to population by the progress of manufactures than by the progress of agriculture. In England during the 18th century the population of the agricultural county of Lincoln remained almost stationary; and the increase in the West Midland Counties did not exceed one-fourth. In the same period, under the influence of the iron trade and the potteries, Warwickshire and Staffordshire doubled their population; and the numbers of the North Midland Counties, which had become great hosiery districts, were also largely, although not in the same degree, increased. In societies as in organisms, growth and development, increased bulk and increased complexity of structure, ought always to proceed with equal pace.

But, however wealth may lighten it, the pressure still remains. Even in the most prosperous community the rate of increase never continues for any length of time unchecked. A large and wealthy society must be considerably developed. Development implies separate occupations. Separate occupations imply skill. Skill pre-supposes not only means and appliances but time. In such a society the standard of comfort is high, and the time required for the attainment of that standard is considerable. The Irish peasant who was content with a bed of straw and a dinner of potatoes, and whose skill consisted in the use of a spade, could marry at eighteen. The professional man whose desires are very different and who consequently seeks a good standing in his profession, must often wait till he is twice that age. It is in fact impossible that a wealthy population should increase without restraint. The two conditions are inconsistent. The same causes which contribute to the increase of wealth, contribute also to the increase of that forbearance which is the main check upon human fecundity. Wealth implies capital; and capital implies foresight, intelligence, and frugality. The development of these qualities thus tends at once to increase wealth and to restrict population. It has been contended* that in proportion to the development of our higher faculties the reproductive powers will gradually become limited; and that the excess of fertility, by calling those faculties into exercise through the difficulties it occasions, produces that enlargement of the nervous centres which is inconsistent with its own continuance. All such speculations however are probably beyond our present reach. Population is year by year increasing; and such an increase, slow though it be, must at some remote period, if it continue, fill the world. But we need not dread the approach of such a period, and we cannot foresee the means by which this tendency will be controlled. The preventive check is sufficient to retard almost indefinitely the question of over-population; and when that question does present itself, the people to whom it will come will doubtless be enabled to deal with it.

§ 6. We have seen that in the case of the lower organisms, if the presence of any check be lightened even in a slight degree, the influence of the geometrical rate of increase will immediately make itself felt. The same principle applies to man. If from any cause the prudential check in any community be weakened, the numbers soon show a remarkable increase. This result takes place quite irrespective of the utility of the consequence. It may be that the standard of population was artificially depressed; and then the removal of the unnatural repression is beneficial. If on the other hand the usual motives to self-control were

---

* See *Westminster Review*, vol. i. p. 499.

by some interference weakened or taken away, the consequent increase of population will be injurious. There are in recent times three remarkable examples of this principle. The first is France after the revolution; the second is England under the old Poor Law; and the third is Ireland before the famine.

In France the abolition of the Feudal grievances and the other proceedings of the revolution raised the great bulk of the population from a state of intense privation to a state of at least comparative comfort. Population therefore rapidly increased, partly because more children were born than before; and partly because of these so born the improved circumstances of their parents enabled an unusually large number to be reared. Thus notwithstanding the destructive wars of that period the population of France rose from twenty-four millions in 1780, to twenty-nine millions in 1815. But when the unnatural repression of the old régime had passed away, the influence of the prudential check began to operate. The succeeding generation grew up with different habits and a higher standard of comfort; and the annual increase of the population, although the country is more prosperous than it has hitherto been, is very slow.*

In England about the latter part of the eighteenth century, the allowance system, as it was called, was introduced into the administration of the Poor Law. According to this system, every labourer whose wages fell below a certain standard, received from his parish in addition to his earnings a sum sufficient in the estimation of the magistrates to keep him in comfort. Thus the man's income was proportioned not to his services, but to his wants; and not to his personal wants only but to those of his family. This system therefore both removed the natural check to population, and

---

* Mill's *Political Economy*, vol. i. p. 416.

offered an actual bounty to prolific unions. No man then needed to consider before he incurred the responsibilities of married life his probable means for fulfilling those responsibilities. Every child that was born was regarded not as an additional charge, but as an immediate and direct source of wealth. The idle man with a large family was better off than the industrious man who had none. Accordingly the population of England nearly doubled itself in half a century. In 1783 it was eight millions; in 1832 it was fourteen millions. Although in this period the same terrible war which desolated France made havoc, to a somewhat less extent, of Englishmen, other and contrary causes probably assisted to increase the English population. But after every allowance has been made, a formidable residue is still left to illustrate the results of legislative tampering with the natural laws of human increase.

In 1652 the estimated population of Ireland was eight hundred and fifty thousand. A century afterwards, it was nearly two and a half millions. In 1841 it exceeded eight millions. The excessive poverty of the people during this period is described with a melancholy uniformity by all inquirers, from Sir William Petty down to the Devon Commission of 1844. These two conditions, inconsistent as they at first appear, are related as cause and effect. But it was the poverty that produced the population, not the population the poverty. It was in the thinly inhabited, and not in the populous, districts, in Galway and Mayo, and not in Armagh or Down, that the poverty was extreme. Yet in the former counties there were but 170 to 180 inhabitants to the square mile, while in the latter each square mile supported upwards of 450 people. How this result was produced, how population, or its absence, tends to wealth or to poverty, I have already attempted to explain. The explanation of the other proposition, the tendency of poverty to bring with it under certain conditions a large population, is equally obvious. There was nothing to which the Irish peasant during the two preceding centuries had to look. Nothing that, at least in ordinary circumstances, he could do would better his condition. His sole available resource was his land; and if he improved his land, he paid as the reward of his industry a higher rent. Why should such a man curb his natural inclination, and abstain from marriage? His church encouraged him; his affectionate nature prompted him; prudence was a quality of whose advantages he had no means of forming a conception. A cabin and a potato garden were all the wealth he knew, or could ever hope to attain. These possessions were generally procurable; and in some cases for political purposes were eagerly pressed upon him. Thus the habit of early marriages became almost national; and the population, freed from all but physical and positive restraints, continued to increase until the Autumn of 1846 introduced its eventful changes.

§ 7. In all the changes through which every seed passes until it becomes a plant, and through which every ovum passes until it becomes an animal, certain characteristic phenomena are invariably observed. The homogeneous and formless germ undergoes a series of differentiations and integrations. The number of its parts is increased, and the structure of each part continually becomes more distinct than before. The differentiations give an advance in heterogenity; the integrations give an advance in definity. At the same time, the numerous and well-defined parts thus distinguished become mutually dependent, and are closely united as components of one whole. A series of changes similar to those which present themselves in organic evolution takes place in society. Most of these changes have been noticed in the preceding pages. It has there been shown how society gradually assumes a co-operative character ; and how, amongst a number of families having similar occupations, differentiations spontaneously arise. These differentiated groups combine so as to form a considerable community; and these communities are, as the process of integration proceeds, ultimately fused into nations. As in the embryo, its general structure, and each separate organ, experience in their respective evolutions similar changes ; so in the social organism, these changes take place not only in its entirety but in its several parts. We have seen how industry as it advances not only divides its occupations, but by a succession of integrative changes localizes each of them ; and how not only each occupation, but each district and even each nation, gradually assumes a special industrial character. Thus, as Mr. Spencer* observes, "Beginning with a barbarous tribe, almost if not quite homogeneous in the functions of its members, the progress has been, and still is, towards an economic aggregation of the whole human race ; growing ever more heterogeneous in respect of the separate functions assumed by separate nations, the separate functions assumed by the local sections of each nation, the separate functions assumed by the many kinds of makers and traders in each town, and the separate functions assumed by the workers united in producing each commodity."

§ 8. This principle, most fruitful in its consequences, is amply illustrated in Mr. Spencer's profound and instructive work. In addition however to the more obvious industrial cases that I have already mentioned and to the illustrations given by Mr. Spencer, a few historical examples may be cited. These examples will both serve to illustrate to those who are unfamiliar with the conception the law of

---

* *First Principles*, p. 161.

evolution, and will at the same time themselves derive from it a new and clearer significance.

In an early state of society where composition in prose and writing are alike unknown, and neither positive records nor scientific culture have yet found place, and where consequently no critical standard of belief is formed, an unlettered but imaginative and believing people naturally desire some explanation of the marvels within them and around them, of their familiar customs and institutions, of their ancestors and of the nature and disposition of those super-human powers of whose influence they were dimly sensible. Hence at every time and in every quarter of the world where the conditions were fulfilled, there have sprung up collections of current stories which profess to supply the desired information. Among the lively inhabitants of the hills of Hellas, these stories, themselves often of surpassing beauty and expressed in immortal song, formed for many generations the exclusive intellectual sustenance of the people. "They are," says Mr. Grote,* "the common root of all those different ramifications into which the mental activity of the Greeks subsequently diverged, containing as it were the preface and germ of the positive history, and philosophy, the dogmatic theology, and the professed romance which we shall hereafter trace each in its separate development." The poet was not in those ages what we understand by the term. He was at once the religious teacher, the historian, and the philosopher, of his day; its novelist, its musician, and its song-writer. But with the advance of civilization this assemblage of confused functions began to present some points of difference, and to acquire specific organs. The philosophers first emerged: then the historians became distinguishable. Gradually the

---

* *History of Greece*, vol. i. p. 46).

religious teaching separated itself from the poetic traditions, and the poets chose their subjects among later events and present interests.

Thus the course of development was for so far complete. From the homogeneous bards with their vague and ill-defined functions, there arose a number of separate classes, each distinctly marked out from the rest. In each of these different classes a similar process may be traced. If we compare the cosmogony of Thales with the science of the present day, what else do we find but the continual repetition of the same principles. New branches of knowledge have been from time to time separated, each of which in their turn are reduced into smaller and more specific divisions, until there seems some danger of our losing sight of their mutual coherence. We know the complexity of the historian's materials, the different aspects from which history may be written, and the number of persons whose co-operation is required, for even such histories as in the present state of knowledge can be produced. Still more obvious are the divisions which abound among the writers of avowed fiction. But the process of development in the case of the religious teachers is the most remarkable of all. In the early ages of Europe, the Church had the exclusive control of all ecclesiastical and almost all civil transactions.* Every form of literary or scientific knowledge was confined to the clergy. Of education they had the exclusive possession. They were the conveyancers, the advocates, and the judges, of the time. The churches and the monasteries afforded the best scope for the talents of the tonsured architects, and decorators, and musicians. Under the protection of these latter establishments, agriculture revived. None but churchmen had the power of recording the events of the time, or

---

* Milman's *History of Latin Christianity*, vol. vi. p. 360.

of collecting any information respecting the past. None but churchmen could prepare any public document, or were competent to conduct the administration of public affairs. But if we follow their history, we may see how by degrees separate occupations have started up : how ecclesiastics have been pushed from, or have resigned, one function after another; until in our own days the last of the great occupations that lingered under clerical control, that of education, has almost entirely escaped from their hands; and the clergyman is constantly tending to become exclusively the teacher of religion.

Again, recent investigations have shown, by a method and with results very different from those in favour a century ago, that ancient society consisted of an aggregation of families. The family, not the individual, is the unit of the earliest form of society of which any record exists. It is by the gradual disintegration of the family, by increased differentiations between its component parts, by the obliteration of the old lines of descent and the substitution of new lines based on a different principle, that of contiguity, that our present state of development has been attained. " The movement of the progressive societies," says Mr. Maine,* " has been uniform in one respect. Through all its course it has been distinguished by the gradual dissolution of family dependency, and the growth of individual obligation in its place." From the simple despotism of the patriarch, from that homogeneous social state in which all the relations of persons are summed up in the relations of families, we have gradually arrived at that complex state where each member of the family constitutes a separate though kindred unit ; and where all or nearly all the relations of persons arise from the free agreement of individuals.

---

* *Ancient Law*, p 168.

Our own political history also furnishes some remarkable illustrations of development. Subject to the common law, all power of making, interpreting, and executing the laws belongs to the sovereign. This authority was soon understood to be exercisable only with the assistance of the Royal Council. Presently a distinction was made between the legislative and the administrative functions of the Council. By a further differentiation of the latter class, the judicial functions were separated from those of the executive. Each of these organs thus separated soon presented further differences. By a remarkable series both of differentiations and integrations, the Royal Council for legislation was developed into the Houses of Lords and of Commons. From the Aula Regia in its judicial capacity sprung the various Courts at Westminster. From the remaining authority of the King in Council, the Courts of Equity and the Judicial Committee of the Privy Council are descended; and traces of the original system still exist in the ultimate jurisdiction of the House of Lords. In executive matters the Royal prerogative is now exercised by many different officers, each perfectly distinct, yet all more or less related. Our whole constitutional history in short consists of a description of the passage of our institutions from their primitive homogeneous form into their present highly complex and diversified condition.

§ 9. The inter-dependence of the parts of any organism is a portion of its evolution. In the undeveloped state a single organ performs a great variety of functions. These functions are essential to the life of the creature ; and development does not supersede but improve their performance. Evolution implies the formation of definite and specific organs, and their subordination as well between each other as to the whole: It follows therefore that the well-being of each part is essential to the well-being of the whole; that the general condition of the whole reacts upon the condition of each part; that each part depends upon the due activity of every other part, and that any injury done to one part must in a greater or less degree affect all the rest. These consequences apply to social as well as to organic evolution. Thus we see the importance of the part that co-operation and exchange bear in the social system. These great agencies not merely increase the powers of industry, but form the very cement and bond of society itself. By their aid men's interests and their duties are inseparably intertwined. No class can entirely monopolize its gains; no class has exclusively to bear its losses. If the miners of Staffordshire strike, 10,000 potters wholly unconnected with the dispute are thrown out of employment. If the supply of cotton be cut off, or if the potato crop fail, it is not merely the persons directly dependent upon cotton and potatoes that are afflicted: but business of all kind languishes; the demand for every description of service is checked, and every class throughout the country feels the calamity. On the other hand a good harvest brings with it not only increased wealth to those concerned in land, but increased activity in every description of business. The crops of Odessa or of Ohio determine the marriage of many an English girl, and regulate the demand for all the comforts and all the luxuries that many an English household requires. The rents of the peer are well paid, the wages of the artizan are large, when there is an abundant wool-clip on the Murrumbidgee, or when the miners at Ballarat have luck. Abundance is indeed the parent, not less than the offspring, of exchange. If it be produced by the action of exchange, it in its turn enables more persons than before to buy, and disposes more persons than before to sell. It is the manifest interest of every person that the industry of every other person with whom he deals should be largely remunerative. It is in a wealthy, and not in a poor locality, that the greatest fortunes are made. It is with wealthy, and not with poor, customers that the sellers of commodities or of services usually like to deal. It is with wealthy and and civilized nations, and not with tribes of naked savages, that we carry on our most profitable commerce. Notwithstanding all that has been said of the passion for money-getting and its hostility to the fine arts, it is not among paupers and barbarians that painters and sculptors, poets and men of science, chiefly flourish. It is only where industry is largely productive, where, after provision has been made for our primary wants, a surplus is found to remain, that men can afford to indulge in luxuries, whether of sense or of intellect, and that consequently other men can find a sufficient inducement to devote their time to the providing those luxuries. Whether therefore we regard individuals, or classes, or nations, their separate prosperity tends to the common prosperity; and no injury can be done to one which will not re-act to the detriment of all. Any attempt to increase the well-being of one at the expense of the well-being of another, although it may be successful in its immediate object, produces on the community results more or less pernicious.

Nor is there any exception to this great principle in the case of those persons whose occupation consists in mitigating those natural imperfections that are incident to humanity. It is in the poorest and least civilized countries that disease is most formidable, that vice is most rampant, that violence and fraud are most frequent, that the difficulties of communication are most formidable; but it is not in such countries that the physician or the teacher, the lawyer or the carrier, who sought only their own gain, would desire to practise. If human nature were so constituted that none of these obstacles to our enjoyments existed, if we had no disease and no vice, if our laws were universally known and were undoubted in their application and were implicitly obeyed, if we could by a mere wish transport ourselves and our commodities whither we pleased, we could dispense with the occupations which depend upon these obstacles; and we should manifestly be by so much richer than we at present are. But as at all times and apart from any exercise of our will, people must be cured, and youths instructed, and rights ascertained, and goods conveyed, it is the interest of those who render such services that their constituents should be so wealthy as to induce them frequently to seek and liberally to remunerate professional assistance; and it is the interest of those who receive such services that the persons who render them should be as skilful as the existing state of knowledge will admit. It is only where the obstacles on which these occupations depend are removable, and are upheld by the vested interests or the prejudice of those whose business it is to remove them, that the interests of any of these classes become adverse to the interests of the public. Yet even in such cases both justice and policy demand the gentlest and most conciliating treatment. "Envy," observes Bentham,* "is never more at ease than when it is able to conceal itself under the mask of the public good; but the public good only demands the reform of useless places ; it does not demand the misery of the individuals holding the place reformed."

---

* *Works*, voL i. p. 320.

CHAPTER XXII.

OF THE ASSISTANCE RENDERED TO INDUSTRY BY GOVERNMENT.

§ 1. In the preceding pages I have assumed that men are free to direct their labour towards any object they may desire; to use such means of assisting their labour as their circumstances or their judgment may suggest; to enjoy exclusively the result of such labour; to transfer to others this exclusive privilege; and to make other binding agreements with their fellows. These assumptions involve the existence of law and of government. They imply the recognition of certain rights, and the exercise of a power to enforce such rights. It now remains to consider the grounds on which these assumptions rest; to trace the nature, and the influence upon industry, of men's political organization; and to estimate the inconvenience which these agencies in practice not unfrequently occasion.

Man is not merely a gregarious, but a political, animal. His tendency towards association is an attribute, not so much of his physical, as of his moral constitution. In the lower animals the impulse to aggregation is always the same. Their experience teaches them no improvement in its use. If their combination be excessive, its pressure does not induce their dispersion. But man's sociality continually increases with every advance that he makes in civilization. The individuality of the labourer is preserved; and at the same time the co-operation of his efforts with those of others is secured. With every extension of society the reconciliation of these conditions becomes more complete. It appears partially in the family; but it involves for its completion further and other relations. Although the family is the social unit, and although it may in some sense be taken as the type of society, yet the relation of the family is essentially distinct from the relation of the state. They differ not in size merely, but in character. The state is indeed formed of an aggregation of families; but this aggregation necessarily gives rise to a new kind of relation. The relation between separate families is not the same relation as that between members of the same family.

The fundamental idea of the family is love; the fundamental idea of the state is justice. The family is a union in which the members constitute one whole with common interests. The state is an association of separate and independent families. When these households combine, the conditions of their associated action must be determined. Out of this combination and for the determination of these conditions the sense of right is evolved. Each man claims the exercise of all his faculties; each man is willing to respect in others the right which he claims for himself. Thus these separate rights are mutually limitary. Every man is entitled to the free use of all his faculties, but only so far as this use does not infringe upon his neighbor's corresponding right. But men are selfish and violent, and are thus easily led to transgress an acknowledged duty. In matters of contract they are often incapable of foreseeing the consequences of their acts, or of precisely expressing their meaning. Some means must be provided for ascertaining and enforcing rights; for taking care that every man shall have the amount of freedom to which he is entitled, and that no man shall have more than that amount. The less therefore that a community is restrained by morality, or is enlightened by general knowledge, the greater will be its need of positive law and of a vigorous executive.

§ 2. The political organization of society seems to be the earliest stage of its evolution. Without such an organization it would be scarcely possible to bring together any number of men; much less to keep them in association. For in man the centrifugal forces are hardly less powerful than the centripetal. Difference, dispute, anger, hatred, revenge, present a formidable series of disruptive agencies. These anti-social tendencies have also a peculiar characteristic. Physical evil may by proper means and sufficient skill be avoided or repaired. Experience and the wisdom that it brings can cure much folly or excess. In either case the removal of the cause is generally attended by the removal of the consequence. But in human discord one conflict springs immediately from another. Wrong begets revenge, and revenge is a new wrong. Some external force is requisite to break through this vicious circle. Some retaliatory act must be found that excludes all further revenge; or some final and authoritative settlement of the dispute must be obtained. This prevailing force is found in that spirit of reverence which is implanted in man, and which is always most powerful when the social tendencies are weakest. It is to sentiment and not to any calculation of utility that government owes its origin. The Russian serf loves his Father the Czar. The Chinese worship their Emperor as the Son of Heaven.* The devotion of the clansman, even the loyalty of the Englishman, are something

---

* Davis's *China*, vol. i. p. 264.

different from any cold consideration of personal interest. Thus a differentiation is produced in the previously undifferentiated mass of families; and the process of organization begins. When it has commenced, the mutual wants of the people and the exchanges thence arising give to the combination consistency and character. This industrial development acts as it were as the cement of the social structure. Accordingly where it is absent, the association quickly loosens and falls asunder. Thus some of the Indian halfcastes are said to display in their hunting and their military expeditions considerable associative ability; but when the immediate object of their union is accomplished, and they have no longer need for each other's services, they immediately relapse into their barbarous isolation. When however the union is permanent, the law of evolution operates. The organ becomes more marked; the function is more precise; and its performance is more accurate.

What that function is, there is little room for doubt. We may hesitate to pronounce it the sole function; but its primary place has never been denied. Before the development of civilization and its many aids to our gratification, society existed. Its object at that time must have been not the interests that were subsequently developed, but the protection of men in the exercise of their faculties; or in other words the maintenance of rights. But the function by which a thing begins to exist is its essential function. Consequently the maintenance of rights is the primary duty of government.* Amid all the disputes respecting the functions of government, there has never been any difference as to the existence of this great first duty. Protection was the object for which the vassal pledged his faith; for which the the client was bound to his

---

* See Spencer's *Social Statics*, p. 253.

patron; for which the clansmen or the leudes gave their devotion to their chief or to their lord. No government even in its lowest state has ever denied that it owed to its subjects protection. No writer upon government ever maintained that this duty did not attach to the sovereign. This function of government therefore may be taken to be universally recognized.

§ 3. The evolution of industry, and indeed its existence, pre-suppose some amount of political organization. The industrial organization springs, as we have seen, from co-operation and exchange. But neither co-operation nor exchange, except in their merely rudimentary forms under domestic control, are possible without some recognition of rights; and rights require some machinery for their enforcement. Even industry in its simplest form, the mere exercise of the individual's faculties for the satisfaction of his wants, requires, as a condition precedent, protection. Still more emphatically does this principle apply to the accumulation of capital, and to its investment for the purposes of invention. If there be any considerable risk that his labour will benefit others and not himself, man will limit his production to the extent that is indispensable to him. In such circumstances, he will not need, and he will not seek, the aid of any of the usual auxiliaries to labour. Consequently we never find any settled industry, far less any considerable amount of industrial development, in any country where anarchy prevails. On the contrary the weakness of government is of itself sufficient to destroy the industry of the most flourishing community, or to neutralize the most ample natural advantages. For examples of the former we have the long catalogue of miseries consequent upon the fall of Imperial Rome. The Republics of South America furnish abundant illustrations of the latter. Industry then is practicable only where there is security; and security can be obtained only where there is government. This is the one great benefit that sovereigns and legislatures can confer upon industry. They have no higher favour to bestow than to afford to every man some tolerable security that he shall enjoy the fruits of his own exertions, his own ingenuity, and his own forbearance. No amount of care devoted by the legislator to any other object will compensate for a deficiency in this essential respect. "In legislation," says Bentham,* "the most important object is security. If no direct laws are made respecting subsistence, this object will be neglected by no one. But if there are no laws respecting security, it will be useless to have made laws respecting subsistence. Command production, command cultivation, you will have done nothing; but secure to the cultivator the fruits of his labour, and you most probably have done enough."

We must not however restrict too narrowly the meaning of this security or protection. For the attainment of our industrial advantages something more than the mere preservation of life and safety of property is required. The state is a jural society.† Its object consequently is not the protection merely of life and property, but the maintenance of rights. Rights denote the claims of each member of the state to the unimpeded exercise of all his faculties, subject only to the similar claims of those amongst whom he lives. In maintaining rights therefore the state has to afford security both to person and to property; but it has to do something more. It must determine what constitutes property, and what amount of interest a proprietor may take. It must regulate the formation of contracts between the living, and the transmission of the interests of the dead.

---

* *Works*, vol. i. p. 303.
† See Lieber's *Political Ethics*, p. 156.

It must determine relations which have been left indeterminate, or not sufficiently determined, by the parties themselves. All those matters which are incidental to the better discharge of these duties come within the limit of state duty. Hence arises all that pre-appointed evidence which forms so important a portion of our law. For the prevention of disputes and for a security against fraud, the law provides in relation to many of the most important transactions of life certain forms and solemnities as the only evidence which it will recognize of these transactions ; and it further establishes certain machinery for the recordation of this evidence.

The industrial importance of this portion of state duty can hardly be overstated. Contracts are the very life of co-operation and of exchange. Yet the contract is of comparatively recent date in the history of laws. The structure of archaic society, and the feelings thence generated were altogether opposed to such transactions. The members of the same family were incapable of contracting with each other ; and the family was not bound by the engagements of its members. When two separate families had any transactions, the proceeding was encumbered with numerous formalities, the omission of any one of which was fatal. Far from the present rules of law being checks upon men's natural and unrestricted power of contract, it is to the law that the contract, as we understand it, owes its existence.* The facilities for the formation of the contract, the security for its performance, and the completion of its deficiencies are all matters essential to industry, and are all directly due to the action of the state. Equally important and equally dependent upon the action of the state are the laws which regulate landed property. , " No improvements," says Mr. Mill,†

---

* See Maine's *Ancient Law*, p. 312.
† *Political Economy*, vol. i. p. 226.

"operate more directly upon the productiveness of labour than those in the tenure of farms and in the laws relating to landed property. The breaking up of entails, the cheapening of the transfer of property, and whatever else promotes the natural tendency of land in a system of freedom to pass out of hands which can make little of it into those which can make more; the substitution of long leases for tenancy at will, and of any tolerable system of tenancy whatever for the wretched cottier system; above all the acquisition of a permanent interest in the soil by the cultivators of it, all these things are as real and some of them as great improvements in production as the invention of the spinning-jenny or the steam-engine."

§ 4. The maintenance of rights forms at all times and in all circumstances the primary duty of the state. But in the earlier period of society this function is, as we might anticipate, less clearly defined than it subsequently becomes. There are in such cases other demands for the action of the state hardly less imperious than those of protection itself. "Without plenty of government," says Mr. Wakefield,* "the settlement of a waste country is barbarous and miserable work." The intervention of the law is indeed more urgently and more constantly needed in the construction of such a society than in its preservation. Men's rights have not been ascertained; custom has not yet been established; and the action of positive law is required to determine the new relations to which new circumstances give rise. Hence it is that, as Mr. Wakefield observes, the quantity of government in such cases is more important than its quality. Some law is necessary; if appropriate, it is well; if not appropriate, the rule is at least fixed and men adapt themselves to its requirements.

---

* *Art of Colonization*, p. 211.

But there are in a new country other duties of transcendent importance which, although they do not involve strict jural relations, the state is bound to perform. Lord Durham remarks that there is no difference in the machinery of government in the old and the new world that strikes a European more forcibly than the apparently undue importance which the business of constructing public works appears to occupy in American legislation. In every part of that country whether under British or American rule, and in every other country in similar circumstances, the construction of public works is the test of administrative merit, the incessant subject of legislative discussion, and the principle item in the account of public expenditure. To older communities that have passed, although very slowly, through the corresponding period of national existence, such discussions seem unworthy of the dignity of a legislature. Yet however different from their own duties these proceedings may be, they are in the circumstances no less necessary and therefore no less suitable. "The defence of an important fortress," says Lord Durham,* "or the maintenance of a sufficient army or navy in exposed spots, is not a matter of more common concern to the European than is the construction of the great communications to the American settler; and the state very naturally takes on itself the making of the works which are matters of concern to all alike."

The same writer also points out that in young communities a still more momentous concern than even the construction and the maintenance of the means of communication is found in the disposal of the waste lands. That these lands are the property of the state or of the sovereign as representing the state, and can be converted into private property under legal regulation and by no other means, has never been disputed.

---

* *Report on Canada*, p. 63.

The state in effect is, in addition to its ordinary duty, a proprietor; and is obliged to assume the duties incident to that character. " Upon the manner in which this business is conducted," says Lord Durham, "it may almost be said that everything else depends." If the acquisition of land in whatever quantities or for whatever purposes be made difficult or troublesome or be subjected to any needless uncertainty or delay, applicants are irritated, settlement is hindered, immigration to the colony is discouraged, while emigration from it is promoted. If, on the other hand, land be granted with careless profusion, the population is dispersed; great tracts of desert land increase the natural difficulty of communications; and obstacles are interposed to co-operation and to exchange, and so to the natural evolution of society. If moreover the public surveys be inaccurate, and the boundaries of the alienated land be incorrectly or insufficiently defined, mischievous litigation must for many years ensue, and the titles to property must be rendered uncertain and insecure. In every new country therefore the construction of the means of communication and the disposal and management of the waste lands form part of the duties of the state. The efficient performance of these no less than of the other functions of government is essential to the industrial evolution of the community. Without them there can be no combination for municipal or other public purposes, no growth of towns, no means of public worship or of secular instruction, no spread of news, no acquisition of common knowledge, no intercourse even for amusement. These advantages and many more than these depend, at least in the earlier settlement of the country, upon its government.

§ 5. We may expect to find, in conformity with the law of evolution, the earlier stages of the political organization marked by that want of distinctness which is characteristic of imperfect development. We may also expect to find the several parts of the organism becoming at each advance more defined and more coherent. Accordingly in the earlier stages of society the functions of the sovereign, and the objects respecting which he exercises them, are very numerous. He is at once father, chief, leader, judge, and priest. All the territory and all the natural agents which it contains are at his disposal. All the improvements made thereon must be made at his command and under his direction. He exercises his authority on all subjects with which the head of the family deals. These subjects include not merely the mutual relations of the members of the family; but also their individual conduct, their dress and their manners, their affections and their beliefs, their amusements and their occupations. Whether the family was expanded into a village, or villages were combined into a nation, this patriarchal system still prevailed.

Out of this primitive and undefined condition of society, where all property was held in undivided possession and administered by an irresponsible authority, individual rights slowly emerged. As the individual became distinct from the family, so the authority of the common patriarch was lessened. The paternal form of government gradually changed. The priestly functions of the sovereign were assumed by a separate class; the laborious duty of administering justice devolved upon officers appointed for the purpose. The extension of private property no longer necessitated a reference to the universal proprietor for every improvement upon the land. Alienations became more facile : contracts were enforced : bequests were sanctioned. Even commercial transactions gradually work themselves free. None but jural relations are ultimately left for the cognizance of the state.

This undeveloped political organization characterizes the Chinese monarchy and the other Eastern despotisms. Such is the condition in which the government of the Czar now rests. Such was in a more modified form the original of the Teutonic monarchies. Even in those settlements which men of advanced civilization have founded, the proper functions of government are extended far more widely than in the parent country. But even in England, the highest existing type of social development, the duties actually undertaken by the state are very multifarious; and the limits of its interference are by no means settled. Of those who concede that there are many matters with which notwithstanding many precedents to the contrary the state ought not to interfere, but at the same time contend for an enlarged sphere of state activity, no one has succeeded in defining the point at which that activity should cease. Even so high an authority as Mr. Mill can suggest no more precise a test than general expediency. But the rule of expediency is not a solution of the problem: it is at most merely a statement of it. The most absurd interferences of the state are due to the belief that they will be beneficial. Even if we knew what things were expedient, we should not have made any approach to determining whether these things should be done by the state. The expediency of the act, or rather our belief in its expediency, is implied in the act itself. When we ask what things it is expedient for us to do, it is a somewhat oracular reply to tell us that it is expedient for us to do those things which are expedient.

A preferable explanation may perhaps be obtained from a consideration of the nature of the state. The state is a society; that is to say, its objects are not expressly stated, like those of a trading company or other association, but arise out of the relations subsisting between its members. But every man enters society, or at least continues in it, in the hope that he will thereby secure greater happiness than he otherwise could do. The happiness therefore of its members, or in other words the gratification of all the faculties of each of them, forms the object of society. All the acts of society consequently should promote the happiness of each of its members, and not of some only to the exclusion of others. Thus the care of those matters and those only which are of common interest constitutes the proper function of government. If all the members of the community be directly interested in any act, the state should undertake the performance of that act. If a portion only of the community be concerned, the matter should be left to the attention of the parties interested in its accomplishment. Such in the main seems to have been the opinion of our greatest political philosopher, although even his words perhaps admit of an undue latitude. "The clearest line of distinction," says Burke,* "which I could draw, whilst I had my chalk to draw any line, was this, that the state ought to confine itself to what regards the state, or the creatures of the state, namely, the exterior establishment of its religion; its magistracy; its revenue; its military force by sea and land; the corporations that owe their existence to its fiat; in a word to every thing that is *truly and properly* public, to the public peace, to the public safety, to the public order, and the public prosperity."

§ 6. There are some apparent anomalies in legislation which this principle explains. Matters which in one country are obviously unfit subjects for legislation excite in another country the greatest public attention. For this difference we may now give an answer somewhat more

---

* Works, vol. v. p. 210.

precise than the usual observation that the circumstances differ. In Holland, for example, the conservation of the sea-dykes is obviously a matter not of individual but of general concern. It does not affect individuals merely, or even large classes of the population; but the lives and property of the whole nation are involved. In Australia the rapid growth of the thistle and the burr, and the facility with which the seed of these plants is diffused, have rendered their destruction a matter of earnest legislative solicitude. The dryness of the Australian climate and the terrible bush-fires that are the scourge of the settlers necessitate very stringent restrictions upon the careless use of out-door fires. The planting with forest trees the vast steppes of Southern Russia with a view to mitigate the aridity of the climate is evidently a matter of public concern. Yet in England neither the maintenance of sea walls nor the extinguishment of fires in the fields, occupies in any peculiar manner the attention of the state; nor does the law interfere for the extirpation of weeds or the planting of trees. In that country none of these matters have any direct influence on the common interest, and they are therefore not objects of the common care.

So too the legitimate legislation, if I may so speak, of the same country, irrespective of the laws that result from ignorance or error, has in different generations very different objects. New matters of common interest arise; and old subjects of importance continue to be such no longer. Before the age of coal and of iron, the preservation of the English oak forests was justly regarded as a matter of national importance. Before the rise of the factory system, it was unnecessary to interfere with the employment of women or children in mills or in mines or in bleach-works. The general safety required, at one period of their history, the prohibition of certain customs and certain modes of dress in Ireland and in Scotland; but no person would now desire a sumptuary law in either of these countries. It is now the common interest of all Australian colonists to preserve the wild animals that are with great pains and at great cost in course of acclimatization in that country ; yet when these creatures are permanently established it is not likely that a system of game laws will be there maintained. In the days of the Plantagenets the crown lands were not less important to the English nation than they now are to Americans or to Australians. The old law-books are full of cases in which the proprietary rights of the crown were concerned. But the colonial practitioner of the present day, upon whose attention grave questions of prerogative are often forced, derives little aid in his inquiries from modern reports. Formerly the sovereign in addition to his political duties was a great landed proprietor. His revenue was mainly derived from his domains. He had consequently to survey, to lease, and to sell his lands, his town lots and his country lots, his woods and his pastures, his fisheries and his mines. He had to collect his rents, and to check and supervise his officers. He had to attend to the various profitable incidents that belonged to the crown, his fines, his reliefs, his escheats, his primer seisins, his aids, his wardships, and his marriages. Gradually however the vast domains of the crown were alienated, and the territorial revenues dwindled away. The military tenures and all their incidents were abolished ; provision was made in a different system for the supply of the Royal Revenue ; and the inconsiderable department of the Woods and Forests is now the sole surviving representative of this once great branch of administration and of law.

It must also be observed that government is not a substitutive but a direct agency. Its action proceeds not from the inefficiency of other agencies but from its own fitness. It is not, for example, because they are beyond the power of unofficial industry, but because they are matters affecting the common interest, that the state undertakes the construction of main roads. Mere absolute capacity is not the test of a function. If mere power were the sole difference between associative enterprise and the action of the state, a sufficiently powerful company might undertake not a few of the functions of the latter, even as the state in some countries competes with individuals in their business. But the provinces of associations and of the state are sufficiently defined by the extent of their respective interests. Since the efficiency of a labourer depends generally upon the intensity of his motive, the person who is directly interested in any matter is usually the fittest person for its transaction. Where one person only is concerned, that person is for the matter in question the most competent manager. Where several persons are associated, they and they only are likely to attend to the business. Where every member of the community is concerned, the collective community, like any smaller association, ought to deal with the case. Particular interests will always receive attention in due course. If government have discharged its duty, if every member of the community have freely the lawful exercise of his rights, private enterprise will always be able to accomplish every object of private concern of which in the existing state of society the accomplishment is desirable. There may be cases in which private enterprises cannot be conducted unless they obtain from the state some special power or privilege, and in which the state is entitled in return for its assistance to impose terms upon the applicants. But these cases do not really involve any departure from the usual functions of the state. They arise from the imperfect performance of the function of legislation; and they indicate the need of a revision of some part of the general law. It is those matters only which affect the whole community that properly belong to the care of the state. These matters government is bound to perform, whether private persons be or be not willing to undertake them. A company might succeed in affording protection to persons and to property; but this is a public function and its performance even in the weakest government is never delegated. A company by an arrangement with government and armed with its authority can collect, and often has collected, the taxes imposed by the state. But no civilized country now farms its revenue. It is needless to add that for the same reason and for other reasons now too well-known to require repetition, government ought to abstain from any interference in matters of merely partial interest. The proposition is exclusive both as to its subject and its object. The state and no other power should undertake those duties and those duties only which concern directly every member of the community.

This principle also enables us to draw a distinct line between the functions of the central government and of the local authorities. As in different countries or among different generations of the same country different objects may affect the common interest, so in different parts of the same country there may be matters which concern all the inhabitants of one part, but have no reference to the inhabitants of any other part. All matters then that are of exclusively local interest are the fit subjects of local control. All those matters in which these separate localities and the rest of the country are alike interested are within the province of the central government. If the central government, whether it be monarchical or democratic, continue to perform the former class of duties after the formation of any local organs capable of performing them, all the evils of retarded development present themselves. If local control be extended to objects of general interest, other and hardly less formidable mischiefs ensue. At the present day among Anglican communities the tendency towards the latter evil is the strongest. The extent to which the excessive activity of the central government has stifled in most European countries the vitality of the provinces has exaggerated in our eyes the worth of the franchises which our ancestors always ranked among the good laws of the Confessor. But examples of excessive local activity are not wanting. We may find them in the old county and baronial courts, which gradually gave way before the extension of the royal jurisdiction; and in later times in the inefficient system of local police. Local bodies, taken collectively, can repair their own roads and construct their own buildings better than the central government can do the same work for them. But their success in this respect does not prove that the administration of justice, and the execution of the law, may safely be entrusted to the officers of the locality. Thus decentralization, when rightly understood, consists in a process of evolution. It apportions among several definite organs the functions originally combined in one. Centralization on the contrary, as the term is usually understood, implies a retrograde movement. It reduces the varied and organized structure to an unorganized and homogeneous mass.

CHAPTER XXIII.

OF THE IMPEDIMENTS PRESENTED TO INDUSTRY BY GOVERNMENT.

§ 1. The impediments to industry that arise from the action of government may be divided into three classes. They may spring either from the non-performance or the imperfect performance by government of its appropriate functions; or from its improper performance of those functions; or from its performance of duties not pertaining to its legitimate sphere of action. As to the first of these classes, little now needs be said. The impediments which it includes are the negatives of the advantages mentioned in the next preceding chapter. Still their influence is not confined to the mere absence of advantages that might have been secured. War and anarchy produce terribly positive results. But the maintenance of rights, including of course protection from violence and fraud, is, as we have seen, a condition precedent to the existence of industry, or at least to the extension of any industrial organization. The execution of this great duty admits of various degrees. Rights are very differently maintained in England, and in Kansas; in the nineteenth century, and in the ninth. According to the approach which the performance of the function makes to perfection is the benefit or the injury to industry. In the savage times of

European anarchy, not a vestige can be discovered for several centuries of any considerable manufacture. Even in the later periods of the middle ages, when government had regained its energy, and civilization had considerably advanced, we read of systematic robberies by men of rank. The castled crag of the German prince became the secure retreat for unsparing marauders. "Robbery, indeed," says Mr. Hallam,[*] "is the constant theme both of the capitularies and of the Anglo-Saxon laws: one has more reason to wonder at the intrepid thirst of lucre which induced a very few merchants to exchange the products of different regions than to ask why no general spirit of commercial activity prevailed."

Such a state of anarchy indeed belongs to an age long before the period of industrial development; and the lordly robbers of Europe have been replaced by a very different class of depredators. But even in our time and in our own country where rights are maintained as never before or nowhere else, examples are but too frequent and too extensive of the imperfect performance of this great function of the state. Equity suits dragging their slow length along for upwards of a century; insolvent land-owners who can neither part with their estates nor profitably manage them; just claims defeated from some technical defect, the conveyance of an estate costing upwards of 30 per cent. of the purchase-money every time that it is transferred, and yet the title never absolutely secure; the insolvent triumphing in his fraudulent gains; the honest witness harassed and insulted; all these and all the other reproaches, present and historical, both of our substantive law and of our mode of procedure, furnish ample evidence of the extent to which the neglect or the inefficiency of the state embarrasses industry.

---

[*] *Middle Ages*, vol. iii. p. 315.

§ 2. The non-feasance of the state, important though it be, is not so fatal to at least the continuance of society as its misfeasance. When government is merely inefficient, men endeavour themselves to supply the want. If there be a deficiency in protection, Lynch-law and vigilance committees furnish their summary procedure and their vigorous executive. Where such remedies cannot be applied, the depredator may be open to conciliation. If the defect be in the administration of justice, men will when it is possible abstain from litigating their rights; and will refer their disputes to some arbitrator of their own selection. If government will not give a good title to the land that adventurers want, they occupy it without any title. If government neglect to construct main roads, traffic limits itself to the dimensions that bush tracks will admit. But although a subject can thus supply the omission or repair the neglect of the state, he is powerless to oppose its hostile action. When the power on which above all he relied for aid is exerted to his detriment, his case is desperate. The insecurity to which the action of the government or its agents gives rise is indeed the only form of insecurity that absolutely paralyzes man's productive powers. Against every other depredator he has some hope of defending himself. Against the force of the society in which he lives, he has none.

The most usual case of governmental malfeasance is found in taxation. The state cannot be maintained without a revenue; and the imposition and collection of a revenue is consequently a legitimate and important function of government. Since the revenue represents the cost to the subject of the services he receives from the state, the due performance of this function can never be of any advantage to industry. The sacrifice we make for the attainment of any object, however cheerfully it is made and however much the object is prized, can not in itself be a good to us. But taxation, although it never can be or produce positive good, may, if it be excessive or ill managed, prove a very formidable evil.

§ 3. The principal injuries of a merely economic nature to which taxation gives rise are the uncertainty of the tax, its excessive amount, and the mode in which it is raised. The first of these injuries is obviously a case of insecurity. If a man never know what is his own, if he be compelled to surrender any portion of his earnings that the officers of the state may think fit, if he has at the same time to endure in its collection all the insolence of office, the worst kind of insecurity is established. Such is the acknowledged cause of the present lamentable condition of many once prosperous countries in the East. Such was the condition of India under Indian rule; and, although the worst European misrule never even approaches Oriental tyranny, such was the principal cause of the poverty of the French peasantry before the revolution. The *taille* or village tax was arbitrarily assessed; was collectively levied as a personal tax; and was subject to perpetual variations. Thus no farmer could foresee in any year the probable amount of his taxation in the succeeding year. "To evade," says M. De Tocqueville,* "this violent and arbitrary taxation, the French peasantry in the midst of the 18th century acted like the Jews in the middle ages. They were ostensibly paupers, even when by chance they were not so in reality. They were afraid to be well off, and not without reason." Each tax-payer had in fact a direct and permanent interest to act as a spy on his neighbor; and to transfer to him a portion of his own liability by informing the collector of the increased wealth which this fortunate neighbor had acquired. To such an extent was

---

* *France before the Revolution*, p. 233.

this practice carried that agricultural societies were unable to offer the prizes which such societies now usually give, lest in the assessment of taxes for the ensuing year the winners should find that they had literally gained a loss.

§ 4. Of the effect of excessive, as distinguished from uncertain, taxation, the history of the later Empire furnishes an instructive example. Within sixty years after the death of Constantine, the great organizer of the imperial financial system, before the barbarian inroads had even approached the frontiers of Italy, not less than one-eighth of the whole surface of the fertile province of Campania was from the operation of that financial system found to be waste and uncultivated land, and as such, exempted from taxation.* This amazing desolation, as Gibbon justly calls it, which could so quickly turn the very garden of Italy into a desert was by no means confined to that country. In the Gallic provinces a land tax to the almost incredible amount of one-third of the net produce of the land was imposed. This ruinous tax, which might in cases of emergency be increased at the discretion of the Prætorian Prefect, and was readjusted every fifteenth year according to the increased or diminished capabilities of the farm, was supplemented by the further exaction of forced labour and a poll-tax. The pressure of these accumulated burdens was continually augmenting. As one tract of land after another was thrown out of cultivation, the tax upon the remainder became continually more oppressive. As increasing poverty diminished the number of those who could contribute the full amount of their poll-tax, the demands on the less indigent were proportionately increased. This result was accelerated by the mode of collection. The curiales, or members of the municipality, were

---

* *Gibbon*, vol. ii. p. 337 (Dr. Smith's Edition).

held to be responsible for the amount of taxes payable within their district. If land were thrown out of culture, they were required to assume all its burdens. If the poll-tax failed to reach the assessed amount, they were obliged to make good the deficiency. Nor were they allowed to escape from their position. Many of them sought for refuge in privileged pursuits; in the church, the public service, the army, or even, it is said, in slavery. But all such modes of escape were forbidden; a rigorous search was always made; and the deserter if found was compelled to return to the calamitous greatness thus thrust upon him.*

§ 5. But even if the sum payable for the expenses of government be precisely ascertained, and be moderate in its amount, the manner in which it is raised may seriously affect industry. In the earlier periods of society, when the relations of the executive to the people were less political than proprietary, the exigencies of the state were supplied by the property of the state. In feudal countries the vast domains of the crown, and the various lucrative branches of the royal prerogative were in the first instance supposed to be applicable to public purposes. It was only after their exhaustion and upon emergency that the voluntary aids, which in England were gradually developed into our system of parliamentary taxation, were, at least in theory, required. Yet it is now well understood that a territorial revenue, although at first sight it appears to cost no person anything, is one of the most costly forms of maintaining its government that any community can adopt. It prevents the proper settlement of the country; and it keeps all the lands which are so applied comparatively, if not wholly, unproductive. A great landowner, as Adam Smith remarks, is seldom a great improver

---

* See Guizot's *History of Civilization*, vol. i. p. 305.

of agriculture; and this remark is *a fortiori* true when that land-owner is the sovereign. But the retention of land by owners who, like the crown, cannot use it as profitably as other persons could do, is a mere waste of the public resources; and consequently retards the industrial evolution. These principles are now generally recognized in practice; and accordingly settlement, and not revenue, is the avowed object of the land policy in most new countries of the present day. For the same reason in such countries territorial endowments are now rarely granted. When it is desired to give public assistance to any object, that assistance is given not in land but in money. As the country becomes fully settled, the population becomes more numerous and more wealthy; and consequently their contributions, as well to the state as for voluntary purposes, soon exceed any direct profits that could be made from the land. The advice therefore of Burke* is now practically followed "to throw the unprofitable landed estates of the crown into the mass of private property, by which they will come through the course of circulation and through the political secretions of the state into our better understood and better ordered revenues."

§ 6. The mode in which the public revenue of most countries is now raised is by indirect taxation. This practice is open to grave objections. So far as these objections are political or administrative, so far as they relate to equality of contribution or to the convenience of collecting those contributions, they are beyond my present purpose. I am not now treating of the art of finance, but only of the impediments which certain financial operations present to industry.

---

* *Works*, vol. iii. p. 367.

It is obvious from the very nature of the case that indirect taxation acts as a disturbing force upon prices. It interferes with the free action either of labour or of some of its aids. It presents an artificial obstacle to the attainment of the desired object; and not unfrequently this obstacle is much greater than the actual amount of the tax would seem to indicate. The spontaneous course of industry is thus disturbed, and various industrial derangements follow. Sometimes industry is diverted into new and less productive channels. We see this result by the return of the trade when the duty has been removed. When coffee for example was taxed at one and sixpence a pound, its consumption was very limited: when the amount of duty was reduced to sevenpence, its consumption rose 750 per cent. When the duty was again raised to one shilling the consumption was again checked. When a reduction to sixpence was effected, the consumption once more largely expanded: and when ultimately all the duties on coffee, irrespective of its place of growth or importation, were assimilated and reduced to threepence, the result was even more remarkable. At the commencement of this century the average consumption of coffee in Great Britain was one ounce per head; at the present time it is about twenty-two ounces per head.* Many persons must therefore at the earlier period have sacrificed their personal inclinations; and the capital required to supply their desires must either have been idle, or have been employed in rendering a less urgent service.

The effects too of these disturbances are often felt in some remote part of the industrial system, and in some unexpected manner. The high duty on coffee was found to check the establishment of coffee-houses, and to promote the sale of intoxicating drinks. In consequence of the excise upon glass

---

* Porter's *Progress of the Nation*, p. 549. Levi *On Taxation*, p. 75.

chemists were unable to procure the vessels suited to some of their purposes; and thus chemical researches were obstructed, and an injury was done to all the arts to which chemistry contributes. The duty on paper through the medium of the jacquard-cards employed was also a heavy tax upon figured silk.* The brick duties prevented the lining of shafts and the tunnelling of workings, and so considerably increased the danger of mining. When the duty on salt was abolished, the manufacture of soda became profitable. The regular and abundant supply of soda altered the manufacture of soap, and extinguished the trades in kelp and in wood-ashes which the previous necessities of the soap-boiler had called into existence. The cost of soap which had thus been indirectly increased was still further raised by an excise duty. Its consumption consequently in 1801 was four times less than 1851, and its place was in some degree taken by injurious and caustic powders. Hence an excessive destruction of clothes was at least to some extent the direct consequence of an excise upon salt. Thus indirect taxation, like every other case of state interference, propagates a multitude of changes, and generally of injurious changes, which previous to their occurrence could never have been anticipated.

Another inconvenience arising from indirect taxation is that the precautions which must be taken to guard against a fraudulent evasion of the duty sometimes preclude, when the subject of the tax is a manufactured article, the possibility of improvement in the process. In the case of the excise upon glass this result was very perceptible. Glass of a peculiar size or glass of a peculiar thickness and glass of a peculiar colour could not be made at all, because the excise regulations under which the manufacture was conducted did not correspond with the properties of the

---

* Spencer's *Essays*, p. 320.

materials employed. Nor was it possible to conduct any experiments for the purposes of ascertaining the causes that affect the properties of the various kinds of glass, or of introducing any novelty into the manufacture. Permission indeed might be given by the excise authorities; but it was always granted with reluctance: and the trouble and annoyance incident to such applications to a public board compared with the contingent character of the advantage to the manufacturer effectually prevented any new investigations.

§ 7. Such are some of the chief obstacles to industry that are caused by taxation. But as the right line is the test both of itself and of the oblique, it may be useful, even if it be not wholly within my limits, to consider briefly the true principles which should regulate the exercise of this great public function. Political society is spontaneously formed for the performance of matters pertaining to those interests that are common to every member of the society. Such a performance necessitates the command of considerable pecuniary resources. The person who derives the advantage in any matter is the person who ought to bear the consequent burden. Since all are alike interested in the action of the state, all should alike contribute to its support. Since all are interested in a like degree, all should contribute in a like proportion. This rule requires not equality of contribution, but equality of sacrifice; not equal rights, but an equal maintenance of rights. Every subject therefore should contribute to the expenses of the state in proportion to his income, that is to his total revenue from whatever source derived. This payment is for value received; and is consequently not an evil in any other sense than that in which every other payment is an evil. It is a sacrifice for the attainment of a much greater good: a sacrifice comparatively trifling indeed, even when it is highest, but still a sacrifice. Nor is it in itself a good, as it has been strangely represented to be. It is a means of obtaining a good, but it is itself something very different. Like every other payment, therefore, it should be made plainly and directly for the object in respect to which it is paid. If the sacrifice be greater, or the return for it be less than is necessary, the remedy is to reduce the one or to increase the other. We only cast an air of mystery over a simple transaction, when we pay a higher price for something else, and get government as it were thrown into the bargain. The contribution for public purposes is a plain duty; and every evasion of this or of any other duty brings with it its own punishment.

On both jural and on industrial principles therefore the proper mode of taxation is an income tax properly levied on every person according to his means.* This method is consistent with natural justice, and causes the least possible interference with industry. What the practical objections to this tax at the present time are I shall not now consider. It is not for me to say whether such a tax ought or ought not to be imposed in any country. Even if its imposition were practically impossible, the principle would not on that account be the less true. But the theory of government which these considerations assume can only prevail in an advanced state of society. They have nothing in common with a paternal or a proprietary government. They assume that the taxpayers regard their contributions not as oppressive exactions but as the reasonable consideration for a highly prized service. In the present condition therefore of our political knowledge and of our political morality, we prob-

---

* See on this subject a paper "On the General Principles of Taxation, as illustrating the advantages of a perfect Income Tax." By W. Neilson Hancock, LL.D. *Journal of Dublin Statistical Society*, vol. i. p. 285.

ably must remain content with a system very far removed from the ideal standard. We can only endeavour from time to time to approach as closely to that standard as our circumstances will admit.

§ 8. The obstacles that the state by its assumption of improper functions causes to industry are so numerous and so varied that brevity in any notice of them is almost unattainable. But I shall confine myself to that undue extension of the functions of the state which is shown both in its regulation of industry, and in its own attempts to conduct industrial enterprises. As to the regulation of industry by the state little now remains to be said. So completely has the true character of this interference been explained that the details of our early industrial history read like the narratives of the examination by the ordeal or of the trials for witchcraft. Yet during several centuries this action of the state was universal throughout Europe: and at the present day it is far from being obsolete. Under this system the state commanded what branch of industry men should pursue and what they should abstain from pursuing. It insisted that the branches of industry which it authorized should be conducted in the manner, with the materials, and at the time, that it prescribed and not otherwise. It fixed the remuneration proper in its estimation for personal services and for commodities. It determined what persons and what classes of persons should engage in each kind of occupation, what time they should spend in learning their business, and what qualifications they should respectively possess. It indicated the persons with whom business might be transacted, and the places at which exclusively negotiations might be carried on. It prescribed the conditions on which men might pass to and fro within the country or to other countries. We now know the real character and effect of these interferences, and of the multitude of similar interferences that so long impeded industry. But the mode in which they operated, and where they still exist continue to operate, may admit of some brief illustration.

The effects of the direct interferences with labour whether complete or partial are obvious. They hinder the labourer from working at the occupation which he desires to pursue ; and they hinder others from employing the persons whom they think proper. But besides checking both directly and indirectly the efficacy of labour, this system obstructed every one of the industrial aids. I shall not go beyond our own statute book for examples. The laws that encouraged one species of industry at the expense of another obviously interfered with the employment of capital. The usury laws had not only this effect, but also checked the extension of credit. When the Reformation Parliament prohibited dry calendering, or the use by dyers of brasil wood, and other innovations that were introduced by aliens, and further insisted that woollen caps and hats should be made " according to the ordinary workmanship before in use," it thereby refused to admit of improvements, or to accept the aid of inventions. The Bubble Act at a later date aimed a formidable blow at co-operation. The combined action of men to obtain the market value for their work was punished as conspiracy. To quote examples of interference with the right of exchange would be to quote the greater part of our earlier statute law, and no inconsiderable portion of much later date. So incessant and so mischievous was this interference in every European community that, if it were not for smuggling, trade must have absolutely perished. "In every quarter and at every moment the hand of government was felt. Duties on importation and duties on exportation: bounties to raise up a losing trade, and taxes to pull down a remunerative one: this branch of industry forbidden, and that branch of industry encouraged; one article of commerce must not be grown because it was grown in the colonies; another article might be grown and bought but not sold again, while a third article might be bought and sold and not leave the country. Thus too we find laws to regulate wages; laws to regulate prices; laws to regulate profits; laws to regulate the interest of money; Customhouse arrangements of the most vexatious kind, aided by a complicated scheme which was well called the sliding scale, a scheme of such perverse ingenuity that the duties constantly varied on the same article, and no man could calculate beforehand what he would have to pay."*

The power of the state has also been so exerted as to check the natural course of social evolution. The policy of the staple towns at which exclusively commodities for export could be sold, the attempts to establish staple trades or in other words to encourage certain forms of industry, the endeavour to retain the old agricultural system against the extension of pasturage, the compulsory culture of some particular product such as hemp, all interfered with the spontaneous differentiation of society. In like manner the statutes † which forbade brewers to make their own barrels, or butchers to tan their own leather, or coachmakers to make their own wheels, tended to perpetuate an excessive co-operation, and prevented the natural integrations that would otherwise have taken place. In short, every attempt to interfere with the ordinary development of a country, to abolish the woollen trade that the linen trade might be encouraged, to charge the manufacturing interest that the farmers may be protected, or the mining interest that the manufacturers may be able to meet the competition of

---

* Buckle's *History of Civilization*, vol. i. p. 255.
† 22 and 23 H. VIII. and 5 Eliz.

foreigners, to refuse in a new country to sell land to the settler for the sake of the grazier, or to exterminate the grazier in the hope of making room for the farmer, to laud some one pursuit, whether agriculture, or pasturage, or commerce, as the especial object of national policy, and to hold out unusual inducements to enter upon that pursuit, or in any way to check the natural course of settlement in any country, tends to produce an unnatural uniformity in the occupations of that country and so to arrest its development and retard its progress.

§ 9. Some of the most serious objections to the quasi-mercantile action of government or its management of industrial enterprises are of a political nature. As the Royal Prerogative in England attained its greatest height when it was brought most frequently and vividly before men's minds in matters affecting its proprietary rights and in its humbler form of business transactions,* so the general authority of government is increased by every extension of its irregular functions. "If," says Mr. Mill,† "the roads, the railways, the banks, the insurance offices, the great joint stock companies, the universities, and the public charities were all of them branches of the government; if in addition the municipal corporations with all that now devolves on them became departments of the central administration, if the employés of all these different enterprises were appointed and paid by the government, and looked to the government for every rise in life, not all the freedom of the press and popular constitution of the legislature would make this or any other country free otherwise than in name."

In a merely economic view the principal damage thus

---

* See Hallam's *Constitutional History*, vol. ii. p. 311.
† *Liberty*, p. 198.

done is the interference with the spontaneous evolution of industry. We have seen that in the ordinary course of events each want is supplied in the order of its intensity; and that the failure of any such supply is consequently evidence that some more pressing requirement is felt. This order is disturbed by the intervention of government. Since the capital at its disposal is obtained from the community, every operation that government undertakes is the displacement of a corresponding portion of industry. If the labour and the capital thus employed had been left to find their natural channels, they would have satisfied some want. If the want were the same as that which the government undertook to supply, the intervention of the latter was needless: if the want were different, that intervention was a disturbance of the natural course of development.

But the same funds are seldom so available in the hands of government as they are when they are in the hands of private persons. This proposition, which indeed is now generally admitted, is established by the fact that almost every undertaking of government is a monopoly, and so depends for its existence upon artificial support. But a monopoly for the benefit of government is not different in its nature from a monopoly for the benefit of an individual. Both are alike subject to the Nemesis that waits upon their kind. Thus the intervention of government not only deranges the natural order in which wants are satisfied; but obtains for a given amount of effort a smaller quantity of that satisfaction which it affords than in ordinary circumstances might be obtained. It has those defects of a weakly and inferior organism which whether in animals or in societies nature removes by death; and it has the further defects which an unnatural and artificial immortality involves.

A still more pernicious consequence of the undue intervention of the state is the check that it gives to the development of practical intelligence and of habits of association. The conversion of the ordinary branches of industry into public functions tends to destroy a natural and most potent agency for public education. Although this tendency involves far wider considerations than those of mere economy, yet its influence upon industry is very serious. Useful as is scholastic instruction, it is insufficient to supply the practical training of the citizen. The business of life forms an essential part of the education of a people. The absence of this discipline, no less than the absence of the usual rudimentary discipline, leaves the mental development stunted and incomplete. In all political affairs the highly-trained German is confessedly inferior to the unlettered Englishman or American. The fatal effects which the absence of political experience exercised on the political thinkers of France in the first revolution are well known.* We are familiar with the want of practical ability frequently observable in men of great speculative powers and of remarkable erudition. It was a remark of the Chancellor Clarendon, no unfriendly critic surely, that "of all mankind none form so bad an estimate of human affairs as churchmen." And this want of political tact and skill is even more conspicuous in that class of learned and studious men whose lives are devoted to academic pursuits. So sensible indeed of this besetting weakness are the most eminent members of that class that, sometimes, as in the case of Savigny, they deliberately seek some share in the duties of active life as a means of intellectual growth and invigoration. But amongst the larger part of every community where Anglican institutions prevail, this practical education is continally going on.

---

* De Tocqueville's *France*, p. 258.

Every part of public business constitutes an adult school, not the less efficient because the students are unconscious of the process. When a man serves on a jury, when he sits as a vestryman or as a municipal councillor, when he acts as member of a committee for any public purpose whether it be for business or for charity or even for mere amusement, he is learning a lesson that no schoolmaster can teach. He learns to think of something else than himself and his immediate affairs ; to comprehend common interests and to manage joint concerns. Such a lesson strengthens the tendency towards cohesion and checks the tendency to disunion. It is besides a discipline both moral and intellectual of the best kind. It gives that feeling of responsibility which a confidence reposed in a man by others rarely fails to produce ; and it stimulates the judgment to a healthy and vigorous action.

All these beneficial results are checked when the state saves its subjects the salutary trouble of attending to public business. " A people," says Mr. Mill,* " among whom there is no habit of spontaneous action for a collective interest, who look habitually to their government to command or prompt them on all matters of joint concern, who expect to have everything done for them except what can be made an affair of mere habit and routine, have their faculties only half developed : their education is defective in one of its most important branches." The longer that this deficiency continues, and the more confirmed the habits to which it has given rise, the greater is the difficulty and the severer the struggle in the return to the natural state. It is essential to the spontaneous development of industry that men should be gradually trained by the means which it provides to their social and political duties. In this as in

---

* *Political Economy*, vol. ii. p. 539.

every other natural process success is the reward of discipline. Before a man can rightly use his freedom, he must be accustomed to be free. Before a man can successfully associate with other men, he must have had some experience of co-operation. Before a man can properly conduct any industrial enterprize he must be acquainted with the wants that he proposes to supply. Slaves on their emancipation generally commit many excesses. When any branch of trade after a protracted course of interference is set free, the reaction is proportionate to the preceding restraint. We wonder at the indolence of the free negro, and are ready to pronounce his emancipation a failure: but we forget the duration of his previous degradation. We lament the extravagance of mercantile speculation in some new business; but the long existence of the old monopoly has escaped our memory, and the effects which such a monopoly produces upon the imagination of those who are excluded from it. Not unfrequently the evil has been so long continued as to render its cessation impossible. Even though the pernicious consequences are plainly seen, the former course must still be pursued. The craving for the false stimulant must be gratified at all hazards. As wrong leads to wrong, so every interference necessitates or at least appears to necessitate its successor.*

§ 10. There is now little risk at least in any Anglican country that the old extravagancies of state-action will be revived. When a reader of the present day learns that about eighty years ago Jeremy Bentham's father was brought before a magistrate on a charge of wearing unlawful buttons,†

---

* See for a remarkable example in the case of the supply of food by government during the Irish famine, *Transactions of the Central Committee of the Society of Friends*, p..22.

† *Bentham's Works*, vol. x. p. 87

he can hardly restrain an incredulous smile. When we read of the pilloryings and the confiscations that in France up to the time of the Revolution punished a deviation from the regulation pattern of soft goods, we seem to listen to a traveller's tales of some mischievous apes, and not a narrative of sad realities acted by civilized men. Nor is it likely that either in the present or any future generation any urgent request will be made to government to carry on under its own direction any great manufacturing or other productive operation. When men are profitably conducting their business, they do not wish to have government either as an agent or as a rival. Yet although its outward manifestations are thus changed, the spirit which gave rise to this perversion of the functions of the state is still vigorous. In one place it appears in the form of communism, or of socialism, or of some kindred system. In another it wears the more modest disguise of protection. Sometimes it speaks in the stern tones of absolutism. Sometimes it sounds the war-note of the modern redresser of every wrong, democracy. Impatient philanthropists can think of no other remedy for the evils which they deplore than a law-made agency. Practical men cannot understand the continuance of the world without their customary support. Men of the most opposite opinions and with the most opposite objects desire to employ each in support of their own purposes the power of the state, but never question the extent of that power or the expediency of its exercise. All are unanimous that there is no human evil which may not be remedied by a proper Act of Parliament

Yet in all such cases, however dissimilar they appear, we recognize the same unbelief and the same presumption. Men have no faith in the existence or in the operation of the natural laws that regulate society; but have full reliance on their own fortune, and their own powers. They are convinced both that the time is out of joint, and that they were born to set it right. "They have so little faith in the laws of things and so much faith in themselves, that were it possible they would chain earth and sun together lest centripetal force should fail. Nothing but a parliament-made agency can be depended on; and only when this infinitely complex humanity 'of ours has been put under their ingenious regulations, and provided for by their supreme intelligence, will the world become what it ought to be! Such in essence is the astounding creed of these creation-menders."*

This strange presumption arises from two defects, the one intellectual the other moral. Of these defects the former proceeds partly from the low state of our political and social knowledge, partly from the indisposition of men to analyze the general terms with which they are familiar, and partly from the common confusion between an acquaintance with the practical details of a process, and an accurate knowledge of the principles upon which it is founded. In these circumstances men fail to observe the facts presented to them, or if they do perceive them to apprehend their true meaning. With the advance and the diffusion of knowledge this unconscious ignorance, and the presumption which it engenders, will gradually disappear.

But this state of mind involves to some extent a moral fault, the removal of which is much less easy. This defect proceeds not from a want of knowledge but a want of faith. It is a mistrust in the natural system by which the world is governed, a doubt that notwithstanding the cogency of the proof the promised sequences of nature will fail. Even in physical science this feeling is not uncommon. In works of construction the attempt to obtain excessive strength

---

* Spencer's *Social Statics*, p. 294.

is not merely wasteful, but, by bringing a needless strain upon the parts least fitted to receive it, is absolutely mischievous. Yet, even where the principles involved are well known, this defect is the one which is most common in new machinery, and which when once it has been practically adopted is most difficult to remove. For centuries English ships were built on a principle which involved great waste of material and great inefficiency, although the mathematical principles that the art of ship-building involved were perfectly understood. "It requires," as it has been well observed,[*] "no common powers of calculation and not a little faith for men to trust to the safety of structures which have apparently been deprived of half their former strength." It is not therefore surprising that in social affairs this hesitation should be strongly felt. Few men have strong convictions on such subjects. Few therefore even of those who have studied the laws of society can forbear from some immediate action, either too impatient or too uncertain of the result to wait. It will indeed be long before these impediments to industrial progress can be completely removed. Yet not on that account should those who believe in the sufficiency of the moral order of the world, and desire to see its action unimpeded by human interference, bate one jot of heart or hope. This period of doubt and of uncertain groping after truth is an inevitable stage in social advancement. In due season it will pass away. A single life-time, though it be of the longest, is but a short time in a nation's history; and although the changes we desire to see may be hidden from our eyes, they are only reserved for a generation better prepared for their enjoyment. There is a passage in Herodotus which Dr. Arnold was fond of quoting expressive of the

---

[*] *Edinburgh Review*, vol. lxxxix. p. 66.

bitter pain that arises from the combination of knowledge and of helplessness. But those who like Arnold zealously battle for the true, the just, have indeed their own mission, but need their own discipline. The reformer must be taught to feel that the good will come not as he wills or when he wills; but easily, almost spontaneously, when the world is fitted to receive it. Yet not the less, though success may be far distant, is he to labour, and so far as in him lies to promote the good work. "The highest truth the wise man sees, he will fearlessly utter; knowing that, let what may come of it, he is thus playing his right part in the world; knowing that if he can effect the change he aims at, well; if not, well also, though not *so* well."*

---

* *First Principles*, p. 123.

CHAPTER XXIV.

OF SOME CAUSES OF POVERTY.

§ 1. Adam Smith, in a passage to which I have already had occasion to refer, observes that every man is rich or poor according to the degree in which he can afford to enjoy the necessaries, the conveniences, and the amusements of human life. In the preceding pages I have attempted to show the circumstances that are conducive to the acquisition of that abundance which according to this description constitutes riches. The causes of riches are the labour of the possessor or of those persons whom he represents; the character and the amount of the natural agents that are at the labourer's disposal and that he is able to use; and the extent of the assistance he obtains from those several agencies which I have included under the common title of industrial aids. But poverty is the absence of riches, and consequently its causes include the absence of the causes of riches. Where labour is bad of its kind, or irregular, or misplaced, where natural agents are scarce, or ineffective, or difficult to manage, where capital is not forthcoming, where inventions have not been introduced, where co-operation is uncertain or inadequate, where exchange is limited and costly, where society is still undeveloped, and few expedients for facilitating intercourse have been adopted, where government is weak to protect, and strong only to harass, its subjects, where these circumstances or any of them are found, poverty, more or less distressing according to the number of such misfortunes and their intensity, will always appear.

But the causes of poverty are not exclusively negative. All the conditions of wealth may in any case exist, or may be capable of existing; and yet there may be forces which prevent or impede their operation. Man may deserve success, but cannot always command it. Forces, whether physical or human, which the labourer can neither foresee nor control, may disturb the most sagacious calculation, and overthrow the most reasonable hopes. Cases also occur where industrial conditions are favourable, but where men are reluctant to make the necessary effort. The desires of the labourer may be feeble, and the cost of their satisfaction, although moderate in the estimation of other men, may to him appear too great. Sometimes this reluctance arises not from any indisposition to labour, but from some association connected either with labour generally, or with some particular form of it. Again although the desires may be vivid, the predominating desire may be for objects not included in riches, and inconsistent with their accumulation. There are also efforts which are successful for their immediate object, but which either tend to destroy the utility of the natural agents by whose instrumentality they have been accomplished, or are injurious to the rights of other men. Such are some of the leading positive causes of poverty; and their consideration so far as it has not been anticipated forms the subject of the present chapter.

§ 2. The disturbing forces which intervene between human efforts and the desires that these efforts were intended to satisfy, may proceed either from some physical influence or from man. The former class includes all casualties by fire and flood, by storm or shipwreck, by landslip or earthquake or volcano, or by blight, drought, disease, or any other inclemency of heaven. Within the last twenty years Montreal and Quebec and San Francisco have been each of them more than once desolated by fire. In its earlier days, a bushfire nearly ruined the half of Victoria. Sudden floods have but recently spread destruction through the Eastern Counties of England. The vines and the mulberries of France have suffered from the plague of insects. A mysterious disease destroyed for several successive years the potato crop, and so altered the whole condition of Ireland. A disease hardly less mysterious is now decimating the herds of Australia. It is needless to tell of the conquests of the sea; or of the triumphs of the hurricane or the avalanche, the earthquake or the volcano. In the presence of these great elemental powers man's helplessness is confessed, and he can but impotently gaze on the ruin that they have made.

It will suffice merely to indicate the calamities that arise from the act of man. The unspeakable horrors of war, of which its waste is the very least, now require unhappily no illustration. The greatest disaster that civilization has yet encountered, the detestable struggle which has rent America, enforces with horrid emphasis its frightful lesson. I have already noticed the impediments to industry that the dishonesty or the violence of fellow-citizens present. The interference of the state is another cause of failure. Apart from any flagrant injustice, its action is generally felt in the attempt to remedy the evils of some previous interference. It is not the least of the misfortunes that attend the irregular activity of the state that its cessation, like the disuse of stimulants, is difficult and painful. However unwise any system of commercial policy may be, society endeavours to adapt itself to it; and interests soon grow up under the artificial system, of which the continuance and the removal are almost equally dangerous. By an act, for example, of the Imperial Parliament in 1843, a large premium was in effect given for the grinding in Canada of American wheat for the British market. The Canadian merchants and millers in reliance upon this enactment made arrangements for carrying on the lucrative trade thus offered to them; and a large portion of the available capital of the colony was invested for this purpose. But these arrangements were hardly completed, and the new mills had but just begun to work, when in 1846 the corn laws were repealed, and the advantage that Canada had received was abolished. A terrible amount of loss was thus occasioned to individuals; and a great derangement was caused in the finances of the colony.*

§ 3. The most common causes of failure are those which are incidental to the organization of society. Where a man supplies his own wants, if the positive conditions of riches be present, and if no disaster from without such as those I have already mentioned befall him, his labour must be successful. But in society the case is very different. Where men undertake by way of anticipation the supply of the future wants of unascertained customers, new and formidable causes of failure present themselves. In such circumstances a man may miscalculate the nature, or the intensity, of the desires of that class or community that he undertakes to serve. He may form an erroneous estimate of his abilities to render the service in question, or of the difficulties of rendering it. He may not have acquired proper information as to the extent to which other persons are prepared to render the same service. Examples of these errors are abundant. The absurd projects that in times of speculative

---

* Earl Grey's *Colonial Policy*, vol. i. p. 220.

manias are started, and still more the extraordinary mercantile ventures that are made when a trade is suddenly thrown open, illustrate the failures that arise from mistaking the character of the want. Warming pans were once sent to British Guiana, where however they were turned to some use in the sugar business. A cargo of skates was one of our first exportations to Rio Janeiro. An enterprising speculator is said to have consigned a lot of horse hair wigs to Caffraria. Sometimes the error is less glaring, but is not less fatal. The peasants of Denmark decorate their dresses with silver buttons and bunches of ribbons of particular shapes and colours. Several commercial travellers sought to tempt the fair Danes with a great variety of beautiful buttons, but without success. One, however, more fortunate or more skilful than the rest, carried with him samples of the very thing both in patterns and in shape; and the delighted shop-keepers were only too happy to order from so intelligent a firm not only their buttons but all their other metal wares.*

Of miscalculations respecting the intensity of desires for objects which in moderate quantities would be in fair demand, gluts and the premature development of industry are examples. The former phenomenon is too common to require any detailed explanation. Like the case last under consideration, it is very frequent at the commencement of a new trade. When the South American trade was first opened, it was not only supplied with unsuitable articles, but useful commodities were sent in such quantities that from the want of warehouses the most costly merchandise was left to perish on the beach. Half a century afterwards, the same scene was witnessed at San Francisco and at Melbourne. The shipments useful and useless of 1853 and

---

* Laing's *Denmark*, p. 381.

1854 were sufficient to have supplied the whole of Australia for the next two or three generations. It became difficult in such circumstances to effect sales at any price, and a disastrous crisis was the result. Even after the first excitement of the gold discoveries had long passed away, notwithstanding repeated assurances that six or at the utmost eight moderate cargoes per month were all that the Californian market would require or could consume, the shippers of the Eastern States persisted during 1856 and the following year in sending monthly from twelve to fifteen ships of the largest size fully laden with assorted merchandise. Notwithstanding the unsatisfactory account of sales, the Eastern speculators seemed unable to comprehend that their cargoes were out of all proportion to the wants of the country. For one venture of 200,000 dollars in dry goods, not a penny was ever returned. At length the great commercial catastrophe of 1858 put an end to this madness, and to many similar delusions.*

The sudden removal of restrictions also leads men to attempt a premature division of employments. Privilege suggests gain. When therefore the privilege is withdrawn, numbers rush to share in the treasure that had been so jealously guarded. Where a dealer proposes to supply a new settlement and consequently a small market, if he confine himself to a single branch of business, either his expenses will absorb his profits; or his prices must exceed the ordinary profits of the place. As the latter alternative would only cause a still greater reduction in his business, his only hope of success consists in the extension of his operations. But if with this object he increase his stock, he will have more goods than he can sell. He must therefore either keep them till they are spoiled, or by his terms induce purchasers to

---

* Evans's *History of the Commercial Crisis*, p. 102.

take commodities that they do not really want. The consequence of the latter operation is unsuccessful speculations and bad debts. Thus whatever course he may adopt, failure is sure to ensue. An instructive example is found in the history of American Banking. Upon the removal of some restrictions upon the trade of banking, many banks were started in the thinly populated States of the South and West. Most of these establishments failed. They were founded in districts where all men desired to borrow money and none had any to lend; and where only small amounts could be profitably used. The banks had thus no local deposits; and in seeking to force a business by the introduction of large amounts of capital from a distance, only brought ruin upon their customers and themselves. "A man," says Mr. Carey,* "who undertakes prematurely to deal in money finds that his neighbors have but little to lend, and that he cannot live by his trade unless he can have a large commission. To increase his business, he obtains a supply of capital from abroad in hopes to make considerable profit by lending it out: but he soon finds that he cannot safely do so; that although he might advantageously place ten thousand dollars, he cannot place fifty thousand without incurring great risk. He has however agreed to pay interest, and if he cannot lend his capital he will be ruined by the operation. He therefore risks it in the hands of those who have no immediate use for it; and the consequence is a rise of prices, or in other words a fall in the value of money. The borrowers commence speculations which end in the ruin of themselves and the money-dealer. Every premature attempt at extending the division of labour is attended by similar results."

Examples of the other cases that are included in this

---

* *Political Economy*, vol. ii. p. 244.

class are abundantly obvious. Their ambition and their confidence in their own good fortune continually drive men to engage in occupations in which they have no reasonable prospect of success. We daily see men undertaking duties for which they have no aptitude, and quickly succumbing in the battle of life. Works suspended for want of funds and contractors ruined by their engagements prove that the cost had not been duly counted before the enterprise was begun. The care and pains which the mercantile class takes to procure trustworthy information as to the quantity of goods likely to come into the market, the numerous mercantile newspapers and circulars and lists and prices current, sufficiently attest the importance that these enterprising and sagacious men attach to a knowledge of the proceedings of their competitors.

When a large amount of capital is seeking investment, and where the ordinary fields of employment are sufficiently occupied, the heedlessness of capitalists, and their impatience for profitable investment, greatly aggravate the tendency to the evils I have mentioned. In such circumstances reasonable speculations are carried to an extravagant excess. Absurd and impracticable projects receive encouragement. Companies are formed sometimes without any other object than to gamble in their shares. The sinking fortunes of decayed firms are supported : new firms flourish by the help of a fictitious credit. Every contrivance for obtaining credit or facilitating the operations of exchange is abused to the utmost. Houses that have long been insolvent have thus been known to maintain their position for years. At length the expedients fail : the evil day can no longer be postponed, and the fact must be avowed that the capital has been lost. Then too that intimate relation of the various parts of the social system to which I have before referred is clearly perceived. One failure brings on another. The pressure commences with a slight shock in some distant place. In 1857 at a time of great apparent prosperity, American railway shares became depressed; then a wealthy corporation in Ohio failed. Within a month afterwards, almost all the banks of the Eastern cities suspended their cash payments. The merchants of London, of Liverpool, and of Glasgow at once felt the pressure. Four mercantile houses in Glasgow failed, owing to a Glasgow bank an aggregate sum of sixteen hundred thousand pounds. The Bank stopped payment: some days previous to that event a bank in Liverpool had closed its doors; the alarm became general; and a commercial crisis of unusual severity set in.*

§ 4. However terrible these failures may be, and from whatever source they may proceed, their ravages are soon repaired. " It is," says M. Guizot,† " with human activity as with the fecundity of the earth : from the time that commotion ceases, it re-appears and makes everything germinate and flourish. With the least glimpse of order and peace man takes hope and with hope goes to work." The strength of this recuperative energy and the shortness of the time that its operations require have been at all times a subject of unfailing surprise. The causes therefore of such a phenomenon may well deserve some short consideration.

There is a series of positive terms with which we are familiar. Production, riches; enjoyment, consumption, such is the ordinary course of events. But there is a corresponding negative series with which happily we are less acquainted. This series presents the ominous terms, failure, poverty, suffering, privation. Yet widely different as is the process in the two cases, the final result is the same. The man who

---

* See *Report of Committee of House of Commons on Bank Act*, 1858.
† *Civilization in Europe*, vol. i. p. 131. (Bohn's Edition.)

spends the entire amount of a casual increase of income is after his expenditure no richer than before. The man who reduces his ordinary expenditure by the entire amount of a casual decrease of his income is after his privation no poorer than before. If the former had saved his money, he would have been so much the richer; he would have had not the enjoyment, but increased means of procuring future enjoyment. If the latter had not retrenched his expenses, he would have been so much poorer; he would have had his present usual enjoyment, but he would have been the less able to procure the means of future enjoyment. "Thus," observes Dr. Longfield,* "wealth arises from deferred consumption, and poverty from deferred privation. Consumption destroys the effect of production, and privation puts an end to the effect of failure. The good or the evil is more or less permanent according as the whole enjoyment or the whole privation is longer deferred."

The destruction therefore produced by famine or by waste or any other calamity leaves no permanent poverty. The greater part of the objects destroyed would have been spent in enjoyment. They are now spent without enjoyment; but in both cases they are spent, and in both cases, though by a different course, things return to their former position. Privation has done the work of enjoyment. If the enjoyment had been deferred, still the consumption would have taken place; and the enjoyment of the new product would at some future time have followed. Whatever be the cause of the destruction of the capital, whether it be failure, or accident, or violence, or wilful waste, it is gone as if it had been used not for capital but for revenue; and its loss must either at once or at some future time be endured. The extent and the continuance of the privation depends upon

---

* *Transactions of Dublin Statistical Society*, vol. i.

the extent and the character of the destruction. If no great reduction in the industrial population have taken place, if they have sufficient instruments to work with and sufficient food to maintain them until the harvest, if the natural agents be unimpaired, if in a word the instruments of production exist, production will proceed, and the traces of the injury will be gradually effaced. In proportion to the extent that any of these conditions are deficient, will be the length and the intensity of the suffering.

We may thus see why a war, at least such a war as barbarian invaders have often waged, is the worst calamity that can befall an industrial country. It combines all the evils of famine, and of pestilence, and of elemental catastrophe. A famine caused by the failure of the crops leaves untouched the means of future production. All the capital that has been sunk in works of permanent utility remains as before. The pain indeed is sharp but it is transient. If the people can by any amount of privation cultivate their ground and subsist until the next year, the suffering will almost if not altogether disappear with a favourable harvest. Pestilence on the other hand, as far as it diminishes the industrial population, inflicts a much more serious and more lasting injury. Yet even here population in favourable circumstances springs forward with surprising power to fill up the void. But the influence of war is far more deadly than the inclemency of the seasons or the dreaded pestilence. Not only, like the former, does it destroy the immediate produce of the year; not only like the latter, does it sweep away the industrial population, the makers of wealth; but like the terrible convulsions of nature, it annihilates those accumulations of preceding industry which facilitate and increase the efficiency of labour. The destruction caused by a volcano or a hurricane or an earth-quake, although complete, is generally local. An invading army may desolate an entire country. It destroys the roads, the railways, the buildings, the reservoirs, the harbors, and all the other permanent improvements upon which the productive power of a nation so largely depends. War involves a loss not only of the floating but of the fixed capital of a country; and as the loss is greater, so the suffering will be longer and more severe.

§ 5. We have already seen that the energy of labour directly depends upon the motives to that labour, upon the probability of success, and the assurance of quiet enjoyment. Yet even where there can be no doubt that the returns to labour would be large, and where a fair degree of security is established, a remarkable indolence or aversion to effort is often perceived. The capacity of desire, like every other capacity of our nature, varies in different individuals; and in the same individual admits of either development or decay. The more it has been exercised, the stronger it becomes; and the less that it has been called into action, the more difficult is its stimulation. When therefore the desires are originally feeble, or where from the facility of gratifying the primary appetites, the largest class of desires has seldom been called into exercise and is consequently undeveloped, the habit of indolence may have been so firmly established as to overpower the first feeble cravings of a desire newly felt and as yet imperfectly comprehended. This state of feeling prevails among men who by the bounty of nature or the accidents of fortune are exempted from the necessity of labouring for their daily bread. It is conspicuous among the emancipated negroes. Unless he be in absolute want, no amount of wages will induce the Jamaican black man to work for hire. His desires are few and simple. For his food and his drink the bounty of nature profusely furnishes the mango, the bread fruit, and the cocoa nut. His other wants can be satisfied at the cost of a few hours' work. The taint of slavery too still clings in the minds of these people to hired labour. They prefer their ease and their sense of freedom to wages and to the hated associations of work done for a master. Accordingly when the sugar planter entreats his negro neighbors to return to the cane fields after ten o'clock in the morning and there earn another shilling, or to work in harvest time more than four days in the week, the usual reply is, " No, tankee massa, me tired now ; me no want more money." " And who," says Mr. Trollope,* " can blame the black man ? He is free to work or free to let it alone. He can live without work, and roll in the sun, and suck oranges, and eat bread fruit ; aye, and ride a horse perhaps, and wear a white waistcoat and plaited shirt on Sundays. Why should he care for the overseer ? I will not dig cane holes for half a crown a day and why should I expect him to do so ? I can live without it ; so can he."

In a famous European community, among men whom we seldom even in thought compare with the negro, we find under the influence of similar causes results curiously resembling the feelings and the conduct of the black Jamaican. The following is the description of the Neapolitans given by Mr. Laing.* " In this country the labouring man is no fool who asks what enjoyment or gratification can high wages gained by constant hard work give me equal to the enjoyment of doing nothing, of basking in the sun or sleeping in the shade doing nothing ? Fuel, clothing, lodging, food, are in this climate supplied almost spontaneously to man. Fuel to cook with is all we need of firing, and even that may be dispensed with by most working people, for our food is sold to us ready cooked at the corner of every street. It would be waste and no comfort in it to light a fire in our own

---

* *West Indies*, p. 65.
† *Notes of a Traveller*, p. 107.

dwellings. Clothing we only want to cover our nakedness. A ragged cloak or sheepskin jacket three generations old does that. Lodging is only necessary to sleep in and shelter us from the rain. A mere shed like a coach house does that. We live out of doors. Animal food is not necessary where olive oil is so plentiful as to be used for frying all vegetable and farinaceous food, and assimilating it as nutritious aliment to flesh' meat. Olive oil, wine, Indian corn, flour, legumes, fruit, are to be got in exchange for our labour at vintage and at harvest during a few weeks when these crops require a great number of hands at once. Why should we labour every day? This is the condition of all around us in our station; why should we labour?"

The same indolence which thus characterizes an entire community, is frequently apparent among ourselves in individual cases. It is not uncommon to see men whose natural abilities are amply sufficient to ensure success wasting, after they have procured a precarious subsistence, their time and powers in idleness or some frivolous pursuit. Examples frequently occur where the prospects in life of a young man are ruined by his possession of an income sufficient for his immediate wants. That which would have been a useful aid to an energetic man, becomes to a less active nature a clog. When his companions are acquiring under the wholesome pressure of necessity habits of energy and of skill, the seeming favourite of fortune relies upon his income; and the time for the formation of these habits slips away. He has as much or nearly as much as those with whom he lives, more than most men of his own standing, and he has not the inevitable expenses which the maintenance of a family entails. He does not therefore care to work, or if he think that his income will not ultimately suffice for his future wants, he postpones the evil day; and indulges, until he knows not how to stop, in his favourite amusement.

§ 6. This weakness of the acquisitive faculties is altogether different from the deliberate preference of comparative poverty. Men often forego opportunities of wealth because its acquisition is inconsistent with other objects of desire. Sometimes they wish to be undisturbed in some favourite enjoyment. Sometimes they wish to devote themselves to some particular object. Sometimes their conscience, sometimes a regard for public opinion and their social position, counteracts their desire for wealth. Such is the case with missionaries and the more zealous teachers of religion or students of secular knowledge; and such too is the case in a greater or less degree with those persons who find in the honour attached to their pursuits a sufficient compensation for small pecuniary rewards. The weakness of the acquisitive faculties implies, as weakness in itself always implies, a defect. The preference of other objects to wealth may imply soundness of judgment and the predominance of the higher parts of our nature. Such implication is not indeed necessary, for the selection is not necessarily good, nor is the prodigal who, with a full knowledge of the consequences and upon a balance of advantages, prefers his enjoyment to thrift, an object for either moral or economic approbation. If in such cases the preponderating desire be merely selfish, the character only excites our contempt, and the life of the man is of no use to any other than himself. If it be otherwise, the character always commands our respect, and the selection that has been made may be of even the very highest benefit to mankind.

A remarkable example of the practical influence of counteracting motives, where the desire for wealth is at the same time very strong, is found in the case of the Jews. This industrious and thrifty race sacrifices to its religious observances not less than one third of the year. Their religion commands them to abstain from labour on Friday evenings, Saturdays, and festivals; the laws of the country in which they reside forbid their working upon Sundays. The consequences of this loss of time are more important than might be at first anticipated. Christian masters are reluctant to take Jewish apprentices. The trades in which the Jew can successfully engage are limited, because he can seldom compete with men whose hours of work considerably exceed his own. Where he can maintain a competition, the Jewish artizan earns less than his Christian rival. Hence the earnings of children become more important in the family of the Jew than of the Christian, and the temptation prematurely to withdraw them from school is increased. Even while they nominally remain at school their attendance is frequently interrupted by their detention to assist in some domestic labour; and consequently the benefit of their training is greatly reduced. Thus the Jewish people, in the aggregate and apart from individual exceptions, will probably always be comparatively poor and ill-educated.*

It has often been observed that in Christian communities the number of holidays, especially in those countries where the festivals of the church are numerous, tends to increase the poverty of the labourer. The effects of such holidays have probably been exaggerated. Mr. Laing† remarks that, if we take into account the time occupied in the performance of the various social duties connected with his religious position and sentiments, the Scotchman gives to his religious concerns many more working hours in the year than the Italian. Yet no person regrets the time that among ourselves is thus spent, or thinks that the development of the religious faculties has any tendency to impoverish the community. The poverty of the Irish peasant has been often attributed

---

* *Journal of Statistical Society*, vol. xxv. p. 511.
† *Notes of a Traveller*, p. 269.

in part to this kind of idleness. It has been estimated that he loses more than one hundred days in every year partly from his observance of church holidays, and partly from his attendance at various places of amusement. But the religious sentiments of the Irish peasant are not less intense and his sympathies are not less warm in other countries than they are in his native land. Yet when his industrial conditions are altered, no complaints are heard of the interference of his other duties or of his pleasures with the ordinary course of his industry. It is not to the strength of the counteracting motives, but to the weakness of the ordinary motives, that in this case and in similar cases the idleness of the people is due.

§ 7. Even when the desires are vigorous and with ordinary diligence might lawfully be amply gratified, it sometimes happens that men are indisposed to make the required efforts. This degradation of industry arises from various causes. Sometimes the motive is religion. In Japan for example, where the religion of the country pronounces unclean all dealers in dead animal matter, the butchers and the tanners and their attendant trades are debarred from all social privileges. No respectable person therefore will engage for any remuneration in these occupations. Sometimes among a warlike people almost every peaceful pursuit is but slightly esteemed. Such was the case in Ancient Rome, where the maxim "*Omnis quæstus patribus indecorus*" by limiting the natural field of employment for capital led to no insignificant political results. Even among the degraded Romans of the present day there prevails a sort of debased imperial pride, a belief that a Roman as such is superior to other men. The native Romans despise all manual labour, but most of all labour for another. They accordingly monopolize all the semi-independent trades of their city. So intense is their preference for this kind of employment that the extent to which small trades are carried on by persons without capital and deeply in debt is said to be one of the most formidable evils at Rome.*

In Mediæval France and Germany a gentleman could not exercise any industrial occupation without losing all the privileges of his rank. Even in England, where feudalism was much less virulent than in other countries, the terms mechanical and base were long synonymous. This low estimate of industry extended even to the higher professions. Had Lord Erskine, as he used often to say, been the son not of an Earl but of a Marquis, he could never have forsaken for the bar his earlier pursuits. Even at the present day we never hear of the son of a peer being a physician or an engineer. So too in Russia, at least before the recent change in that empire, no person engaged in trade could own serfs: and a person who could not own serfs, could not be a proprietor of land. This among other disabilities causes the Russian merchant to feel some sense of degradation while following his business. It becomes an object to him to enter the class of the nobility. When he does so, he must withdraw his capital from trade. The abandonment of his industry is the price that he must pay for a better social position.†

By far the most influential cause of the degradation of industry is caused by slavery. It is not the least among the pernicious effects of that system, that under it the ideas of industry and of servitude are inseparably associated. In some of the West Indian Islands no wages will in the absence of absolute want tempt the free black to work as a hired labourer. In his mind compulsory labour and labour for another person are identical. But when an

---

* *Journal of Statistical Society*, vol. xxiii. p. 237.
† *Reports of Secretaries of Legation*, 1858, p. 128.

adequate motive free from any degrading association was presented to him, when he became a proprietor and was to work on his own land for his own benefit, his industrial capacity proved to be by no means deficient. Still more ruinous to the white man are these degrading associations. They seem indeed almost invariably to produce a peculiar form of social disease. Poverty and ignorance combined with the pride of race and the contempt for the ordinary pursuits of industry do not furnish very healthful conditions for the human mind. In Barbadoes before emancipation the condition of the poor whites is said to have been most lamentable, and even yet the effects of the mischief have not passed away. They had no property: the pride of colour forbade them to work; and they were often forced to subsist upon the charity of the negroes themselves.*

In the Southern States of America this consequence of slavery is especially conspicuous. In these countries there are, or rather at the commencement of the war there were, upwards of five millions of "mean whites," the "poor whites" of Barbadoes, and the "petits blanc" of the French Antilles, too poor to keep slaves and too proud to work. These men disdain to work "like a nigger," and prefer a hard and precarious livelihood in the wilderness to that steady industry to which the slaves around them are compelled. Nor is this feeling exclusively confined to the working classes. Where custom has assigned certain occupations to slaves, employers are as reluctant to have white labourers in these occupations as the white labourers are to engage in them. The employer and the employed alike feel the use of white labour in occupations that are usually servile to be humiliating. But there is a less sentimental reason for this preference of slave labour. The labour of such white men as will work in a slave state

---

* Mr. Merivale's *Colonization and Colonies*, p. 83.

is very inefficient. They will work so long only as their actual wants compel them. They work reluctantly and are impatient of control. Thus their labour is unenergetic and their habits are unsteady, and consequently the amount of supervision which they require is excessive.* Servile labour is not merely itself inefficient, but it blights every other kind of labour that approaches it. This result is part of the Nemesis of slavery.

§ 8. Sometimes the course of industry is such as, although it may meet with temporary success, to cause a permanent deterioration to some natural agents. There are some natural substances such as stones or minerals which, in the circumstances at least in which they are attainable, are exhausted by their actual use: there are others whose powers are continually replaced. It sometimes happens that through improvidence or ignorance agents of the latter class are so used as to destroy, either entirely or partially, their reproductive energy. Some animals, especially those which yield the more highly prized furs, have been nearly exterminated. Since the enormous extension which railways have given to the market for fish has led to increased contrivances for their capture, symptoms of over-fishing have been observed. Quantities of fish too are taken while they are still too young to breed, and still greater numbers are destroyed when they are assembling to spawn. In France these wasteful practices completely desolated the rivers and the bays. This great loss and the demand which in Roman Catholic countries exists for fish led to the re-discovery of the forgotten art of pisciculture. But the most important instance of this class is the waste of the elements of fertility in the soil. In many countries the soil has become deterio-

---

* Olmsted's *Journeys in the Cotton Kingdom*, vol. i. p. 112.

rated by the abstraction from it of those substances which form the food of plants, and the neglect duly to restore them. In parts of France and Spain, and in most of the countries west of the Vistula, the vegetable mould is said to have disappeared, and mere "geest land" remains, the unsubstantial semblance of its former self. The same thriftless exhaustion of the soil may be observed in some parts of Australia. The same crop is grown year after year without intermission and without the application of manure until the land is worn out, and its owner abandons it for virgin soil.

This form of waste is most extensive in new countries. Although in older communities ignorance often occasions irreparable mischief, necessity enforces some attention and some attempt to supply the defect. Where fresh land can be readily and cheaply obtained, there is no such restraint upon the owner. But it is when this facility for obtaining fresh land is combined with slavery that the tendency towards this waste is at its greatest height. The proper cultivation of the soil requires a rotation of crops; and variety of any kind is inconsistent with slave labour. The field-hand can only be taught a single set of operations, and these of the simplest kind. Since superintendence is costly, and since without superintendence no slave will make the slightest exertion, these operations must also be of such a nature that they can be performed in gangs. Whatever crop therefore in any plantation is found most profitable, whether tobacco or cotton or sugar or rice, that crop is cultivated exclusively and without intermission. In a few years of such treatment the richest soil is exhausted. Accordingly in Virginia, in Maryland, in Georgia, in the Carolinas, in Alabama, even in the newly occupied soils of Texas, we read of plantations once abundantly fertile, now unfenced and abandoned, worn out and given up to the vulture and the wolf. On the hill sides of the lower Mississippi, for example, in addition to the unceasing growth of cotton, great quantities of soil are from the careless mode of cultivation constantly washed away into the adjacent swamps. Although the soil is of unusual richness, a number of successive crops varying from ten to twenty suffices to render the plantation an absolute desert. Mr. Olmsted* says that he in one day passed in this district four or five large plantations, the hill sides worn, cleft and channelled like icebergs: stables and negro quarters all abandoned, and everything given up to desolation and decay.

§ 9. But deplorable as is the waste of natural agents, it is insignificant in every sense when it is compared with the waste of man. There are great classes of persons whose efforts for the satisfaction of their wants are constant indeed and energetic, but are made at the expense, or in furtherance of the vices, of other men. I do not now speak of crime in its fearful moral aspects; but even in a mere economic point of view, its cost to the community is appalling. According to recent judicial statistics,† it appears that in England and Wales an army of 135,000 people is known to be maintained by the plunder or the vices of their fellow-citizens, living sometimes in profligate extravagance, at other times in the deepest want, at all times and everywhere in profound moral degradation. It is estimated, and the calculation is probably beneath the truth, that the annual cost to the community of this criminal population, whether it be maintained in prison, or by the exercise of its ordinary pursuits, cannot be less than ten millions sterling.

That in a large community there will always be crime,

---

* *Journeys in the Cotton Kingdom*, vol. ii. p. 150.
† See Levi's *Annals of Legislation*, vol. vii. p. 54.

and that an ignorant and violent population will furnish a large amount of crime; and that the circumstances favourable to the growth of crime are also productive of poverty may be conceded. But it is now ascertained that poverty and distress and ignorance and violence in reality add but little to the great mass of crime. This calamity is mainly due to the existence of a distinct criminal class. Into this class are constantly drawn numbers who fall into crime from drunken idleness and vicious indulgence. But the source from which this horrible profession is recruited is mainly those who are more sinned against than sinning, the uncared for, neglected masses. Squalor and neglect are conditions under which the foul parasites of the social system are engendered; and their existence, at least to any considerable extent, may be taken as evidence either of some deep-seated disease or of culpable and gross neglect. There is amongst us a mass of preventible crime no less than of preventible disease, equally costly, but far more appalling. As a mere pecuniary question the cutting off the supply of crime from England alone would be a national gain exceeding the total amount of income and property tax from the three kingdoms.

The adult criminal must be left to the care of the policeman. His home is the prison or the convict depôt. Improved prison discipline and careful supervision may restore to society many of even these outcasts; and so far as they have been the inheritors of crime, some tardy reparation may thus be made for the neglect which has led them to so terrible a school. But the children, the rising generation of thieves and prostitutes, what is their condition? Because they have been born to the saddest lot that this world can give, because they are not orphaned and destitute, but trained by their parents for crime, are they to be left to such an education; or is it only after they have qualified themselves by crime to receive our charity, that we are willing to help them? Is the Reformatory the only home that we can offer to the child now innocent, sure hereafter, unless heaven in its mercy quickly send death, to become guilty? Such was not the spirit of our Common Law, ever noble and generous in its principles, although from ignorance or error those principles have been often misapplied. That law declares that the Sovereign is "*parens patriæ*," the father of his people, the gracious and merciful ideal of Royalty to whom the lunatic and the orphan, and the worse than orphaned child of the profligate, alike look for comfort and defence. At Common Law when the moral character of the parent is so depraved or his religious sentiments are so perverted or obtuse as to render him an unfit protector for his child, the Chancellor, in whose charge the exercise of this prerogative is placed, may remove the child to some more suitable home. It has been distinctly laid down that this authority extends to all subjects without distinction. In practice however its exercise is strictly limited. The Court of Chancery will not make an order which it has no means to enforce. It cannot enforce the support and education of a child whose means depend upon the will of the parent whose control the order of the court is designed to supersede. The court therefore will not exert its authority except in those cases in which the infant is possessed of property in its own right. Yet this is not a case where one law may be suffered practically to exist for the rich, and another for the poor; nor ought the Sovereign, while she gives her wealthy children bread, be compelled to offer to the necessitous not even a stone. It is the duty of the nation to enable the Sovereign to exercise fully this portion of the prerogative. The children of those who are known to belong to the criminal classes should be removed from their contaminated homes, at the expense of the parents if the obligation can be enforced; at the public expense, when such enforcement is impracticable. By what machinery this object could be effected it is not now necessary to discuss. There could be no difficulty in finding proper courts to exercise in a cheap and summary mode a recognized principle of equitable jurisdiction.

As equal care should be shown for the protection of all classes of orphaned or neglected children, so when they become wards of the state, equal care should be shown in their treatment. These poor children should not be cut off from all domestic ties and congregated together in cold and unloving asylums. Such a course is not adopted towards the wealthier wards of Chancery; and its practical results have invariably disappointed the expectations of its advocates. The social unit is the family; and the conditions and the discipline of the family are essential to furnish the natural education for the child. If parental care be denied to the young, the substitute should resemble the natural shelter as nearly as may be. The humble as well as the wealthy ward of the Crown should be admitted, under the care of some appropriate guardian, to the strong yet gentle control of family ties. The evil effects of rearing children in an unnatural separation from all the charities of home, even though the institutions in which they are placed be well and carefully managed, have been shown in the disastrous results of charity-schools and of workhouse education. On the other hand, in whatever country and with whatever creed the family system has been tried, whether it be among the Presbyterians of Aberdeen, or the Protestant orphans of Dublin, or the Roman Catholics of Nantes and Chateaubriand, its success has been undoubted. Under the old boarding-school system, in Irish Protestant schools, upon an average of ten years about forty per cent. of the children turned out badly: at the present time the loss of the Irish Protestant orphans under the family system amounts to about five per cent. Even the cost of the efficient system is less than that by which the child is spoiled. A workhouse child in Ireland costs about £7 a-year : a Protestant orphan in the same country does not cost more than £6.*

§ 10. These causes of poverty are in their nature temporary. They are in no way essential either to the constitution of man or to society. So far as regards the negative causes of poverty, I have already shown that it is in society alone and in proportion to its development that the conditions of wealth can be attained. Of its positive causes, those influences which most disturb the acquisition of wealth, indolence, ignorance, intemperance, are evanescent evils. In many cases and to a great extent, they have already disappeared : there is no unreasonable ground for hope that hereafter they will be still more effectually banished. They are but the weaknesses of our moral nature, which, as that nature is purified and strengthened, by that very discipline disappear. Nor is there anything in the constitution of society that either itself tends to evil, or that is calculated to generate any such tendency. All those social phenomena which now offend or pain us, admit of an adequate explanation. The evil is in every case evanescent. Sometimes the suffering is due to the miscalculation or mismanagement of the sufferer himself. Where the causes have been beyond his control, it is sometimes caused by the pressure of a period of transition, aggravated by some injudicious interference of the state. Sometimes the weakness or the cupidity of government is in fault. Sometimes with the best intentions a public policy is adopted that interferes with the actions of some industrial aid. Sometimes the

---

* See two Papers by W. Neilson Hancock, LL.D., *Journal of Dublin Statistical Society*, vol. ii. p. 325, vol. iii. p. 13.

state neglects its proper duties and so causes both individual suffering and public loss. And sometimes the children in the third or the fourth generation are visited by the slow but inevitable consequences of their forefathers' misdeeds. But none of these evils form any part of the social system. All or any of them may be removed not only without injury to that system but to its great advantage. Between the morality of the individual citizen and the development of the state there is a constant reaction. The higher the moral development of the individual, the greater is his capacity for social organization: the more complete is the organization of society, the greater are the advantages both material and moral that its citizens will enjoy.

Yet to some persons society, although it may tend to the production of wealth, seems to be by no means equally conducive to well-being. They point to the alarming crimes and the gigantic frauds that from time to time affright the land from its propriety, and maintain that these are either generated by society, or that society is impotent to prevent them. Still it is certain that in an advanced community there is less wrong than exists at an earlier period of its progress. In all uncivilized communities, the penal law occupies exclusively the attention of the lawgiver. The subjects which in such circumstances are included within the protection of that law are very different from those with which our criminal code has to deal. Acts which are now justly regarded as heinous offences were once considered either legitimate or at least venial. Piracy was in early Greece a recognized and honourable occupation. The only form of dishonesty noticed in the ancient Roman law was theft. Our own ancestors established a tariff for homicide. The morality of the present day is manifestly something very different from that which such a state of law represents.

and frauds now gives us is evidence that an opposite course of conduct generally prevails. If the generality of men had not been peaceable and honest, these offenders would have had no opportunity for their iniquities. It is the nature of the human intellect, as Lord Bacon has observed to be affected with what is affirmative and positive rather than with what is negative and privative. We are therefore struck with some glaring but exceptional offence, and the countless multitude of just and honourable dealings remains unnoticed. How numerous such dealings are, and what a tendency there is in an industrial community to their increase is manifest. The general honesty of the community, as a mere matter of business and apart from any moral consideration, is essential to all industrial operations : and the sense of its necessity contributes largely to develope in the community both the feeling itself and the disapprobation with which improper conduct is regarded. So influential does this public opinion become and so salutary is its direction that it is to it, and to the general habits and manners that it produces far more than to any positive law that our present security for person and for property is due.*

Nothing therefore can have less foundation in fact than the doctrines of the virtue of the pre-social man and of the industrial advantages of an early state of society. These doctrines, although proposed at different times by men of very different characters and with very different objects, are yet related. In truth the theory of Ricardo is but the complement of the theory of Rousseau. According to the latter, the formation of society was the cause of all moral ills. According to the former, the advance of society brings with it at least a tendency towards physical privation. Not one of the propositions involved in these theories can now,

---

* See Hallam's *Middle Ages*, vol. iii. p. 158.

so far as our evidence extends, be admitted as a fact. Man was never solitary : he was never without property : the more closely he approaches such a state, the more obtuse are his moral faculties : he never consciously entered society, with or without any agreement: he never in an early period of society has used the most efficient natural agents : he never as such a society advanced was driven by want to the use of inferior agents. On the contrary, society, property, law, security, abundance, are the results of man's nature. When they exist, and in proportion to their influence, his moral faculties receive their natural development. Ample capital, the use of more powerful natural forces, a large population duly organized, and all the advantages of co-operation and exchange constantly extending and growing more elaborate and complete, both attest and accelerate this social advance. In such a movement there are no signs of deterioration, of crime, or of want. Its results are purer morals, better laws, ampler security, and more overflowing abundance.

www.ingramcontent.com/pod-product-compliance
Lightning Source LLC
Chambersburg PA
CBHW051235300426
44114CB00011B/744